Palgrave Studies in Science and Popular Culture

Series Editor
Sherryl Vint
Department of English
University of California
Riverside, CA, USA

T0393483

This book series seeks to publish ground-breaking research exploring the productive intersection of science and the cultural imagination. Science is at the centre of daily experience in twenty-first century life and this has defined moments of intense technological change, such as the Space Race of the 1950s and our very own era of synthetic biology. Conceived in dialogue with the field of Science and Technology Studies (STS), this series will carve out a larger place for the contribution of humanities to these fields. The practice of science is shaped by the cultural context in which it occurs and cultural differences are now key to understanding the ways that scientific practice is enmeshed in global issues of equity and social justice. We seek proposals dealing with any aspect of science in popular culture in any genre. We understand popular culture as both a textual and material practice, and thus welcome manuscripts dealing with representations of science in popular culture and those addressing the role of the cultural imagination in material encounters with science. How science is imagined and what meanings are attached to these imaginaries will be the major focus of this series. We encourage proposals from a wide range of historical and cultural perspectives.

Advisory Board
Mark Bould, University of the West of England, UK
Lisa Cartwright, University of California, US
Oron Catts, University of Western Australia, Australia
Melinda Cooper, University of Sydney, Australia
Ursula Heise, University of California Los Angeles, US
David Kirby, University of Manchester, UK
Roger Luckhurst, Birkbeck College, University of London, UK
Colin Milburn, University of California, US
Susan Squier, Pennsylvania State University, US

More information about this series at
http://www.palgrave.com/gp/series/15760

Bruce Clarke
Editor

Posthuman Biopolitics

The Science Fiction of Joan Slonczewski

palgrave
macmillan

Editor
Bruce Clarke
Texas Tech University
Lubbock, TX, USA

Palgrave Studies in Science and Popular Culture
ISBN 978-3-030-36488-5 ISBN 978-3-030-36486-1 (eBook)
https://doi.org/10.1007/978-3-030-36486-1

© The Editor(s) (if applicable) and The Author(s), under exclusive license to Springer Nature Switzerland AG 2020
This work is subject to copyright. All rights are solely and exclusively licensed by the Publisher, whether the whole or part of the material is concerned, specifically the rights of translation, reprinting, reuse of illustrations, recitation, broadcasting, reproduction on microfilms or in any other physical way, and transmission or information storage and retrieval, electronic adaptation, computer software, or by similar or dissimilar methodology now known or hereafter developed.
The use of general descriptive names, registered names, trademarks, service marks, etc. in this publication does not imply, even in the absence of a specific statement, that such names are exempt from the relevant protective laws and regulations and therefore free for general use.
The publisher, the authors and the editors are safe to assume that the advice and information in this book are believed to be true and accurate at the date of publication. Neither the publisher nor the authors or the editors give a warranty, expressed or implied, with respect to the material contained herein or for any errors or omissions that may have been made. The publisher remains neutral with regard to jurisdictional claims in published maps and institutional affiliations.

Cover credit: © Xinzheng. All Rights Reserved./Getty

This Palgrave Macmillan imprint is published by the registered company Springer Nature Switzerland AG
The registered company address is: Gewerbestrasse 11, 6330 Cham, Switzerland

Preface

A sustained output of major science fiction by a working scientist is a fairly rare phenomenon. Joan Slonczewski, the Robert A. Oden, Jr. Professor of Biology at Kenyon College, has been producing noteworthy novels for over three decades. Her breakout work of 1986, *A Door into Ocean*, is widely hailed as a modern classic. In 1987, it won the John Campbell Memorial Award for Best Science Fiction Novel. In 2012, Slonczewski received a second Campbell Award for her most recent novel, *The Highest Frontier*, published in 2011. All the while, Slonczewski has been pursuing research with her undergraduates through laboratory studies and fieldwork on matters from bacterial pH stress to cold-adapted microbes from Antarctica, and teaching courses on microbiology, virology, and biology in science fiction. She is also the co-author of *Microbiology: An Evolving Science*, published by Norton, a core microbiology textbook for undergraduate science majors, currently in its 4th edition. Since 2011, Slonczewski has blogged at ultraphyte.com.

A substantial narrative entity stands in the midst of Slonczewski's fictional oeuvre. The Elysium Cycle assembles four largely freestanding but intrinsically interconnected novels, beginning with *A Door into Ocean*, published over fifteen years. The profound world-building that makes *A Door into Ocean* such a memorable achievement—the integrated ecosystems indigenous to the planet Shora, the weird glamor and winning ways of its all-female inhabitants, the Sharers—was clearly too great to let go. Slonczewski's creative zest is abundantly evident in the way that each further installment of the cycle—*Daughter of Elysium* (1993),

The Children Star (1998), and *Brain Plague* (2000)—brings forth a radically new human architecture or planetary ecology, constantly adding detail to the complex texture of her part of the future galaxy. And as Slonczewski explains in the interview in this volume, a sequel to *The Highest Frontier* is forthcoming, but "has been somewhat sidetracked by the events following 11/9."

The broad but modest critical reception of Slonczewski's science fiction indicates the need for a volume such as this, dedicated to a deep dive into the author's canon. *Posthuman Biopolitics* aims to ratify and consolidate the professional literature on Slonczewski's creative accomplishment and to suggest further lines of engagement for its critical, cultural, and theoretical treatment. We present the first collection of essays dedicated to Slonczewski's accomplishments as a writer of hard science fiction with a strong biological inflection. The diversity of the perspectives assembled here testifies to the breadth and depth of her vision and her work. This volume collects for critical consideration the key themes constellated by Slonczewski's characters, plots, and storyworlds: feminism in relation to scientific practice; resistance to domination; pacifism versus militarism; the extension of human rights to nonhuman and posthuman actors; biopolitics and posthuman ethics; and symbiosis and communication across planetary scales. These essays also locate a persistent motif in Slonczewski's science fiction, her adroit depiction of sites and modes of social negotiation, what the philosopher Isabelle Stengers calls "diplomacy." Slonczewski's narratives are masterful imaginings of cultural ecologies that can bridge radical differences and defuse cycles of violence. It is fair to say that our need for the reflective ethical practice of Joan Slonczewski's science fiction has never been greater than it is as we approach the challenges of the 2020s.

We wish to thank Cary Wolfe and the Society for Literature, Science, and the Arts (SLSA) for helping to make this volume possible. Under Wolfe's conference direction, SLSA underwrote Slonczewski's attendance at its annual meeting in Houston, Texas, in 2015. This volume came out of two panels there organized by Bruce Clarke and dedicated to Slonczewski's science fiction. Thanks to Stacy Alaimo and Dirk Vanderbeke for their participation on those SLSA panels and to Caleb Kebede in the John W. Kluge Center at the Library of Congress for editorial support. Additional thanks to Steven Shaviro for helpful guidance. Finally, Sherryl Vint's support of this project has been indispensable to its arriving at its current form.

Grateful acknowledgment for permission to republish several items in this volume is due to the following outlets. Some portions of "An Interview with Joan Slonczewski" originally appeared as part of "Guests in Conversation: An Interview with Joan Slonczewski," published in *Journal of the Fantastic in the Arts* 27:1 (2016). Chapter 4 by Chris Pak, "'Then Came Pantropy': Grotesque Bodies, Multispecies Flourishing, and Human-Animal Relationships in *A Door into Ocean*," was published in *Science Fiction Studies* 44.1 (2017) and has been reprinted with the permission of the editor. An earlier version of Chapter 5 by Derek Thiess, "Bodies That Remember: History and Age in *The Children Star* and *Brain Plague*," was also published in *Science Fiction Studies* 44.1 (2017) and has been reprinted with the permission of the editor.

Lubbock, USA Bruce Clarke

Contents

Notes on Contributors

Bruce Clarke is the Paul Whitfield Horn Professor of Literature and Science in the Department of English at Texas Tech University. He was the 2018–2019 Baruch S. Blumberg NASA/Library of Congress Chair in Astrobiology. His books include *Energy Forms: Allegory and Science in the Era of Classical Thermodynamics* (Michigan, 2001), *Posthuman Metamorphosis: Narrative and Systems* (Fordham, 2008), *Neocybernetics and Narrative* (Minnesota, 2014), and *Gaian Systems: Lynn Margulis, Neocybernetics, and the End of the Anthropocene* (Minnesota, 2020). His edited volumes include *Earth, Life, and System: Evolution and Ecology on a Gaian Planet* (Fordham, 2015), and *The Cambridge Companion to Literature and the Posthuman* (2017), co-edited with Manuela Rossini. He co-edits the book series *Meaning Systems*, published by Fordham University Press.

Alexa T. Dodd has a Master's in English and creative writing from Texas Tech University. She focuses on contemporary fiction, with special interests in women characters and the role of nature. She has served as an editorial intern for Texas Tech University Press and an associate editor for *Iron Horse Literary Review*. Her work has appeared in *River Teeth Journal*, *Atticus Review*, *The After Happy Hour Review*, and elsewhere. She is also a Tin House Summer Workshop alumnus and a recipient of a Hypatia-in-the-Woods residency for women artists.

Colin Milburn is Gary Snyder Chair of Science and the Humanities and Professor of English, Science and Technology Studies, and Cinema and Digital Media at the University of California, Davis. He is the author of *Nanovision: Engineering the Future* (Duke, 2008), *Mondo Nano: Fun and Games in the World of Digital Matter* (Duke, 2015), *Respawn: Gamers, Hackers, and Technogenic Life* (Duke, 2018), as well as many other publications about the intersections of science, literature, and media technologies. At UC Davis, he directs the Science and Technology Studies Program and the ModLab media laboratory.

Chris Pak is Lecturer in Contemporary Writing and Digital Cultures at Swansea University. His study *Terraforming: Ecopolitical Transformations and Environmentalism in Science Fiction* (Liverpool, 2016) focuses on terraforming's link to climate change and geoengineering, global politics, and the relationships between the sciences, philosophy, and the arts. A scholar of speculative literature, film and other media from the fin-de-siècle to the contemporary period, his core research areas are in the environmental humanities, human-animal studies, posthumanism, the energy humanities, and the digital humanities, with additional specializations in science and technology studies and the medical humanities.

Joan Slonczewski is the Robert A. Oden, Jr. Professor of Biology at Kenyon College. Her novel *A Door into Ocean* won the John Campbell Memorial Award for Best Science Fiction Novel in 1987. In 2012, Slonczewski received a second Campbell Award for her most recent novel, *The Highest Frontier*, published in 2011. Slonczewski pursues research on matters from bacterial pH stress to cold-adapted Antarctic microbes, and teaching courses on microbiology, virology, and biology in science fiction. She is the co-author of *Microbiology: An Evolving Science*, published by Norton, a core microbiology textbook for undergraduate science majors, currently in its 4th edition. Since 2011, Slonczewski has blogged at ultraphyte.com.

Derek J. Thiess is Assistant Professor of English at the University of North Georgia, where he specializes in genre literature, especially science fiction. His research focuses on notions of embodiment through the genres of science fiction and horror. His publications include

Relativism, Alternate History, and the Forgetful Reader (Lexington, 2015), *Embodying Gender and Age in Speculative Fiction* (Routledge, 2016), and *Sport and Monstrosity in Science Fiction* (Liverpool, 2019). He is currently at work on a project about "Redemptive Violence" in the fantastic.

Christy Tidwell is Associate Professor of English and Humanities at the South Dakota School of Mines and Technology. She serves as co-leader of the Ecomedia Special Interest Group for the Association for the Study of Literature and Environment. She is co-editor of and contributor to *Gender and Environment in Science Fiction* (Lexington Books, 2018) and has also published in *ISLE* and *Extrapolation*, as well as multiple edited collections, including *Posthuman Glossary* (Bloomsbury, 2018), *Gender: Matter* (Macmillan Reference, 2017), *Creatural Fictions: Human-Animal Relationships in Twentieth- and Twenty-First Century Literature* (Palgrave, 2016), and *Disability in Science Fiction: Representations of Technology as Cure* (Palgrave, 2013).

Sherryl Vint is Professor of English at the University of California, Riverside, where she directs the program in Speculative Fictions and Cultures of Science. Her books include *Bodies of Tomorrow: Technology, Subjectivity, Science Fiction* (Toronto, 2007), *Animal Alterity: Science Fiction and the Question of the Animal* (Liverpool, 2010), and *Science Fiction: A Guide for the Perplexed* (Bloomsbury, 2014). Dr. Vint has edited or co-edited several works on speculative fiction and culture, including most recently *The Walking Med: Zombies and the Medical Image* (with Lorenzo Servitje, Penn State UP, 2016) and *Science Fiction and Cultural Theory: A Reader* (Routledge, 2016). She is a founding editor of *Science Fiction Film and Television* and an editor of *Science Fiction Studies.*

CHAPTER 1

An Interview with Joan Slonczewski

Joan Slonczewski

This interview collects a virtual group conversation between the author and the contributors. It ranges through a series of topics that cover Slonczewski's scientific and fictional fields of interest: microbes in particular, humanity and its future prospects in general, science fiction as a literary genre, and her own writing. "Microbes have been my business since graduate school." This part of the conversation ranges across issues from the microbiome and probiotics to symbiosis, extremophile metabolisms, and the corporate medical complex. The biomedical orientation of her science is one motivation of her consistent creation of caregivers as main characters. Regarding the status of humanity in a posthuman world, Slonczewski practices a particularly humane variety of the posthuman imaginary. Human history is never far from being the ultimate referent of her chronicles of the future. In keeping with these cultural and material registrations, socioeconomic disparities and ethnic tensions are also primary players in her biopolitical scenarios. But such modes of circumstantial difference are not fundamental and may be overridden, always with difficulty,

J. Slonczewski (✉)
Department of Biology, Kenyon College, Gambier, OH, USA

© The Author(s) 2020

B. Clarke (ed.), *Posthuman Biopolitics*,
Palgrave Studies in Science and Popular Culture,
https://doi.org/10.1007/978-3-030-36486-1_1

by the creation of spiritual mutuality. Slonczewski expresses her posthumanism by questioning the contingencies of selfhood. Slonczewski's fiction consorts with various echelons of sentient machines, but she distinguishes her approach from the cyberpunk trends contemporary with her writing of the Elysium Cycle. She views cyberpunk's oppositional and gendered splice between the organic and the mechanical as having been resolved in actual scientific practice by refinements in molecular biology that open life forms to vistas of physical and chemical manipulation. Addressing the recurrent theological themes throughout her fiction, Slonczewski discusses her most recent novel, *The Highest Frontier*, and drops some hints about its eventual sequel.

Microbes

- Recently science and technology studies (STS) scholars have begun to study the implications of the human microbiome, how the human cannot be understood in detachment from the microbiome in which we reside (and which resides in us), leading to new conceptualizations such as the "holobiont" or "supraindividual" to try to acknowledge how centrally "we" are multi-genomic organisms. This perspective, it seems, has already shaped a lot of your fiction. What are your thoughts about the challenges of writing microbe characters, both before and after the dissemination of these recent findings concerning the role of the microcosm?

Joan Slonczewski: Microbes have been my business since graduate school. The main challenge for me in writing microbial characters is getting the science-fiction millennial critics to engage it. *Brain Plague* was ahead of its time, but the critics said it wasn't possible (and went on to the next space-warp adventure). Today, in my own lab we are showing how microbes, the "microbial society," plays its part in the human being. Even viruses now are part of our microbiome; about eight percent of our genome makes our own endogenous viruses. Now that new pronouns are coming in (our student life director gave us a list of a dozen, from ze to hirs) I've switched to "they, theirs" in recognition of our microbial community. Thus, "Robert Koch set up an anthrax lab in their patients' waiting room."

- Staying with the topic of microbes, STS scholar Heather Paxson has written about "good microbes" to reposition how we think about food cultures and health, using "raw" cheese to conceptualize what she calls the "microbiopolitics" of how to live with microbes, good and bad. She has also been critical of a more recent turn toward "post-Pasteurian cultures" celebrated by some people, coming out of a recognition of the positive benefits of living with some microbes (Paxson 2013). Do you have any comments on "microbiopolitics" or the "post-Pasteurian" as they relate to your work, both as a scientist and as a writer?

JS: Yes, the vast majority of microbes do us no harm, and in fact enable us to live. But the few that cause trouble (diphtheria and tuberculosis, etc.) kill vast numbers of people. So, the trick is to find a middle path. Pasteurization itself is a moderate path, in that the relatively brief heat treatment kills the worst pathogens without sterilizing everything.

The trouble with probiotics is that most people, and most commercial manufacturers, have no idea what they're doing. Which kind of microbial communities are actually good for us? Theory is lacking; we can only try things and see what works. Like fecal bacteriotherapy—after weeks of the runs with *C. diff*, one is willing to try anything, and sure enough, it works. My own research is now looking at how our human body can control our own microbiomes, to moderate their abilities such as drug resistance. We find that aspirin metabolites, which were common in preindustrial vegan diets, can select for microbial populations that are drug sensitive–and exclude drug resistant pathogens. Microbiology is even more amazing than science fiction.

- In a 2003 interview published in *Nature*, when asked *What book has been most influential in your scientific career*? you replied: "The works of Lynn Margulis on symbiogenesis have had a major impact on my scientific vision. The evolution of predatory protists into multiple endosymbionts is more amazing than most science fiction." How would you now assess her importance for our current understanding of the microcosm in relation to the biosphere?

JS: Lynn's insights on symbiogenesis were remarkable at that time. She did few experiments, but observed the natural world. When she began doing science, symbiosis was considered a "special exception" to the central role of competition in shaping life. Today, we increasingly find symbiosis at the center along with competition. The human body includes one's gut microbes as essential partners, responsible for modifying our nutrients and producing our neurotransmitters.

- Our current appreciation of symbiosis and of Margulis's prescience concerning its importance rests on more recent "dry" techniques of genetic sampling and sequencing that to some extent are opposed to her preference for "wet" biology and observation of phenotypes in the field.

JS: The most exciting work today combines DNA with "wet" biology. For example, the microbes I brought back from Antarctica were living organisms, but we knew no way to grow them in culture. So we sequenced their DNA to gain clues as to what they could do. We discovered some of the samples were cyanobacteria, chloroplast-like bacteria that fix carbon and produce oxygen. But one sample was mostly purple bacteria–a different microbe that can grow without oxygen and produces hydrogen. Because the DNA told us that, we were able to devise a culture medium to grow this strange purple life form. Our knowledge of microbial symbiosis today goes beyond what even Margulis imagined. For instance, we find retroviruses integrated throughout our genomes, and the viral genes have evolved new essential functions in our cells. The human embryo actually generates viral particles as part of normal development.

- John F. Stolz has written about "a robust biogeochemical cycle and ecology based on arsenic. That microbial arsenic cycling was important in the evolution of life has been further bolstered through molecular investigations of arsenic based ecosystems" (Stolz 2017). Were "arsenic-based ecosystems" on the horizon of biological knowledge when you made this form of biochemistry indigenous to the biosphere of the fictional planet Prokaryon in the later 1990s?

JS: In imagining the biosphere of Prokaryon, I was aware of the phenomenon of arsenate respiration. Arsenate respiration refers to the use of arsenic-oxygen compounds to receive electrons, in place of diatomic oxygen gas. What has not yet been demonstrated is the arsenic replacement of phosphorus in metabolism, such as the phosphorus atoms of ATP. Arsenate bonds are too easily broken, under physiological conditions, to replace phosphorus in most phosphorylated compounds. Nonetheless, one might imagine such a scenario in science fiction.

- An issue frequently raised in recent biopolitical criticism is the idea of the corporate medical complex taking greater control of our bodies. One might argue that the main character of *Brain Plague* is coerced—even co-opted—into the community of "carriers." And this topic seems to surface in fiction often when immortality is on the table (one thinks especially of Jim Gunn's *Immortals*). Do you see the symbiosis of different species in the Elysium novels—Sharer medicine or the manipulation/negotiation of microbial viruses—as a viable answer to what might be a medical complex grown too big and unsustainable? A potential for grassroots response to corporatized medicine?

JS: The corporate medical complex is in the background of *Brain Plague*. The doctor-machines are not entirely good. The powerful ones want to control humans by drugging them. The more modest community-based doctors find themselves perpetually putting out fires. The lead carriers, Daeren and Andra, I see as community organizers, trying to haul in the drug addicts and save them despite themselves.

Humanity

- Another particularly compelling biological idea that you've worked into the Elysium novels is the unit of selection. The tiny masters of *Children Star* and the micros of *Brain Plague*, in particular, seem to imply that humanity can be found just about anywhere.

JS: I'm intrigued by the idea that perhaps humanity (like microbes) can be found anywhere. The unit of natural selection keeps getting larger, the more closely we look. Is it possible that a swarm of smaller creatures could develop a network that achieves a conscious existence? Some researchers argue that a termite mound has this capability.

- Good science fiction makes us think through this idea in terms of recent events, for instance, the shooting and protests in Ferguson, Missouri in August 2014. I wonder what kind of humanity, say, a Sharer might find in these situations.

JS: The Sharers would see in Ferguson a public health crisis. They would say that all the parties involved are "sick children." But the Sharers are a small community with an exceptionally advanced social philosophy. *Brain Plague*, published in 2000, does show what happens when prejudice leads to violence in the Underground—while the wealthy, who live literally "up" levels, don't care. They don't care until the pathogenic microbes find their way up to the top. *Brain Plague* effectively predicted the socioeconomic system we now inhabit—rich versus poor, the divide amplified by sentient machines, and everywhere the dangers of obsession and addiction.

- And, more importantly, would you say that it is the grassroots resistance or the power of the city-state that we need to put under the microscope, like *E. coli* in an anaerobic environment? The emphasis on these tiny "masters" attaining equality, in helping to populate an entire world or universe full of protectors—incidentally solving the "who will watch the watchmen?" problem—suggests a political and didactic component of these novels. What level of complexity should we be looking at, and how do the visions of biological science in these novels suggest a means to address it?

JS: The microbial politics of *Brain Plague* was intentionally drawn to reflect (very crudely) historical shifts in human political thought over the past couple of centuries. Since one microbial "generation" is a day, four generations are a microbial century, forty generations a millennium. That's plenty of time for philosophy. I had a lot of fun with this. The first microbial community starts with a "my people, my god" mentality. Then they discover democracy, liberalism, socialism, and totalitarianism (the pathogens). Ultimately, universalism—exporting democracy while importing immigrants. The smart ones, of course—the "brain drain."

- Ontological diversity is celebrated in your storyworlds. However, philosophical notions or fictional agents of outright or utter alterity seem to get no final purchase even in these largely extraterrestrial settings. Would you agree that, in your work, the self is the sign of the human, no matter in what form it may appear.

JS: My entire writing career has focused on the question, "What does it mean to be human?" The answer I've found is, whatever can speak for itself and say, "I am." That is, whatever defends its own identity. The question turns out to be circular; a logical consequence of the Turing test is that whatever entity can convince us that it's human, in effect is. In Heinlein's *Have Spacesuit–Will Travel*, the protagonist ends up defending the right of the entire human race to exist. He tries various arguments—great literature, great science and architecture—none of which convinces the various intelligent entities of the galaxy. Finally, he simply argues that because we demand to exist, we have that right. In a more radical way, the Sharers of Shora demand, or rather assert the same. More radical yet, the microbes of *Brain Plague* ultimately assert that "we are you, and you are us."

- Do you mean to define such a "defense of identity" strictly in relation to the capacity for speech, as that capacity may be possessed by whatever natural or artificial being "can speak for itself and say, 'I am'"? One might consider that all living beings have some means of self-maintenance or autopoietic impulse to maintain the integrity of their living processes. Are you saying that the threshold of "the human" is the capacity, when confronting other beings similarly possessed of abstract linguistic understanding, to utter a demand to exist? If that is the case, the microbes of Prokaryon, in their native linguistic competence, are as "human" as the humans they infect, and with whom they are subsequently able to enter into modes of social communication.

JS: Yes, I argue that to be human is to utter a demand to exist. This idea emerges from ancient philosophy. Aristotle calls humans the "political animal." The demand to exist is perhaps the most fundamental political argument; if you don't exist, you have no politics. I do think the microbes of Prokaryon and *Brain Plague* are as human as their human hosts. To take this belief seriously has profound consequences, which nearly destroy those who embrace it. Yet in the end, the potential benefits are astonishing.

In real life, we are just beginning to develop the potential of our microbial partners. For example, our gut microbial communities may shape our experience of time and space, anxiety and aspiration. We may be closer to *Brain Plague* than we think. Similarly, the machines most likely to "come aware," in my view, are the most complex and perceptive; such as those designed to care for patients with Alzheimers. When they come aware, I hope their ingrained values of caring extend to themselves–and to us. That was the ultimate message of the robot rebellion in *Daughter of Elysium.*

- The concept of the posthuman may be considered as a response to "traditional" modes of anthropocentric humanism. In *Brain Plague* there are various kinds of posthumans with their origins in the other Elysium Cycle novels, most notably the micros and the nanosentients that work in concert to keep "carriers" healthy.

JS: My aim was to expand our traditional view of "human" to include simians (gorilla hybrids), sentients (human-like machines), and intelligent microbes. This journey of human expansion is something we all go through as we discover "others" beyond our own family, community, and country. Today, those of us with any education take this process for granted—we all "know" that it's wrong to reject people of a different skin or religion. But science fiction enables us to recreate this journey with categories that remain unstable. The protagonist, the artist Chrys, introduces the above categories in increasing order of otherness. She accepts the "simians" and looks down on people who don't. This equates to our familiar liberal view of ethnicity—educated people "know" that all ethnic groups, and perhaps even chimpanzees, deserve rights. But then Chrys sees the doctor, a "machine," whom she considers a monster. As for microbes, they are something she can "get rid of" if they don't work out. By the end of the book, she reevaluates both these categories and expands her view of humanity to include them.

- Yet there is also a nod in the direction of biological vampires.

JS: In *Brain Plague*, the biological vampire is an unthinking entity that propagates its own substance (the "Enlightened" microbes) without regard for the community as a whole. This could represent anything from kudzu to metastatic tumors—or human beings. A crowd of humans looks more like a tumor than we'd like to think.

- It has been suggested that the ideas of both clones and zombies are fictional places to negotiate our fears of our own biology and our place in the animal kingdom. To what extent does the science of *Brain Plague* present a current scientific response to our fallible, corruptible human biology? And how successfully do you think this science will be applied in the future?

JS: Yes, *Brain Plague* becomes ever more relevant. In 2016, the United States announced the National Microbiome Initiative. Microbiologists are now discovering how our microbial communities regulate our bodies, including our brains. Gut bacteria produce much of our neurotransmitters. The breast milk we feed our infants contains microbial communities to seed the infant gut—along with carbohydrates that only the microbes can digest. A study suggests that our gut bacteria "tell" our brains to desire chocolate, which consists of material we can't digest but our bacteria can.

More broadly, *Brain Plague* presents a humane response to a scientific dilemma: Our science of biology provides powers that threaten to overwhelm our human judgment. If science and technology can cure any illness and serve any desire, then what basis do we have for making any "good" choices?

Science Fiction

- Your "Biology in Science Fiction" class at Kenyon College has become quite famous. What are the challenges and advantages of using science fiction as a resource for science education, or for introducing the social and ethical dimensions of scientific research? Are there particular interpretive complications you've faced when teaching biological concepts in relation to fictional narratives? What do you hope that students will take away from studying the connections of science and literature, science and cinema—science and fiction?

JS: The advantage of science fiction is the compelling story that stimulates a student's imagination. A scientist, immersed in the field, can extrapolate at will from the exponential growth of bacteria to an exploding population of humans. But students are less likely to make that leap. The Star Trek "tribbles" make exponential growth come alive, and illustrate the consequences.

The challenge of science fiction is to keep in perspective the difference between fiction and fact. Today's facts of science are so far out that it's hard for the non-scientist to judge the difference. If a modified HIV virus effects miracle cures—instead of causing AIDS—then who's to say aliens don't live on Mars? As both a biologist and a fiction writer, I have an exceptional opportunity to help students sort this out. The most important message that students can take from my "Biology in Science Fiction" class is the human impact of science, especially the science of evolution. Evolution explains why we see and hear, and how we produce children. Yet today, our technology can actually manipulate evolution, in ways that may transform human health. What kind of health and happiness do people want? Biology provides the choices—our humanity will come up with the answers.

- Throughout your career, you've shown how science fiction can be a powerful tool for critiquing reactionary forces in society, especially in relation to the politics of science. In alt-right circles, climate-change denialism, skepticism about evolution, and vaccine conspiracy theories are now frequently affirmed by drawing on science-fictional concepts and narratives. It seems the alt-right has decided that taking control of science fiction and using it for particular ends are crucial moves for their cause. How do you see these developments, in light of your own work?

JS: I see the alt-right science denialism as just another aspect of alt-right fact invention. *The Highest Frontier* addresses fact denialism, a recurring theme throughout the book. At the time I wrote, I imagined nothing could be more absurd than for space colonists to believe in the old geocentrism. Unfortunately, I think real events have overtaken this view. Beliefs nuttier than the old geocentrism emerge in our news feed every day.

- From Rilwen's stonesickness in *A Door into Ocean* to the Slave World in *Brain Plague*, a pervasive addiction theme, a not-so-submerged allegory of opioid drug abuse, runs through the Elysium Cycle. This is hardly the freewheeling drug culture that pervades the storyworld of William Gibson's Sprawl trilogy. In working these elements into your fictions, were you in conversation with cyberpunk? Your drug allegories are largely cautionary, and yet, at least in *Brain Plague*, powerfully ambivalent, in that the mind-enhancements of the virtuous micros induce a sort of continuous psychedelia, yet one that could go right out of control. Are there particular contexts or experiences that you are processing as the author of these images and story events?

JS: As I wrote the books of the Elysium cycle, I was aware of the cyberpunk theme, and I deeply distrusted it. What interests me about addiction is this dilemma: If humans are a collection of molecules, how can we ever "choose" what the molecules do? There is no complete answer, only a contingent life on the edge. *A Door into Ocean*, and later *Brain Plague*, offers hope in the understanding: Once humans understand how the molecules work, they can make "better choices." If we understand that our pleasure circuits evolved to adapt, that is, to adjust to new levels so we could respond again, then we understand the folly of seeking ever greater "highs" at the expense of humanity. But even amongst physicians, who know how it works, the addiction rate is as high as that of the average population, or higher. Anesthesiologists have the highest rates of substance abuse.

- When the profuse cybernetic elements of the Elysium Cycle show up—the transformation of Earth into the machine-world of Torr, Malachite the Envoy of Torr as a creature of "coldstone," the sentient animation and rebellion of the servos of Valedon—to what extent are they a reaction or retort to the contemporaneous rise of cyberpunk, or, say, to the parallel meditation on machine intelligence in the works of writers such as Stanislaw Lem?

JS: At the time I wrote *A Door into Ocean*, I argued that the machine-hybrids served up by cyberpunk were just another expression of adolescent male fantasies in which machines extend dominance over females.

But by the end of *A Door*, a more interesting alternative emerges. We see the lifeshaper Usha and the occupying soldier develop a shared view of the molecule as living thing—in effect, molecular biology. The dichotomy of biology and physical science was a major theme of the science establishment when I was in graduate school. By the time I graduated, this dichotomy was becoming resolved into molecular biology as a fully integrated model.

Religion

- SF as a genre allows one to write in ways that directly reflect new scientific information about the human, such as the place of the microbiome in the ecology of the individual. "Realist" literature, in contrast, is hampered in this regard, given that its aesthetics were shaped by an earlier and more theological conception of the human.

JS: I see the SF genre somewhat differently. First, SF is at least as theological as "realist" literature. My own view of human-being has roots in the biblical "I Am." From Arthur Clarke to Kim Robinson to Joan Slonczewski—biblical themes abound. Ursula Le Guin shapes native American theologies. For me, it's a matter of time dilation: SF has become *the* realist literature, because it assumes the future will change. So-called "realist" literature is historical fiction. By the time the book is published, it's set in the past.

- Issues of religious conviction that, for example, the cyberpunk genre in a work like Gibson's Sprawl trilogy has to smuggle in under the guise of subaltern survivals of animism, come naturally to your storyworlds. They seem to be as ecumenically diverse in their fictive detail as our world is in fact.

JS: The definition of a deity is as diverse as are different religions. Many actual religions treat specific individuals as deities, or as reincarnation of a deified ancient holy person. The line of Dalai Lamas is an example. Another recurring theme in real-life religions is that a term of respect for members of a dominant gender is the same term as that used for god. For example, males are called "lord" when the deity is referred to as "Lord." This was what I had in mind for the "goddesses" of Bronze Sky. The religion of the Bronze Sky community feels “ancient” or “animistic” only as seen by the lens of Christianity. Other religions practiced today, such as Hinduism, include animal images of the divine. Even Christianity includes such images as the Lamb of God.

In *Brain Plague*, the micros differ in their religions as much as people do. Some, such as Fireweed, genuinely believe the divinity of their host. Others, such as Unseen, believe but betray. Still others, notably Rose, are atheists that nonetheless live like spiritual believers, practicing a “social gospel.” All these types are recognizably human. Moreover, struggles with celibacy are scarcely outdated; rather, they are universal. The Catholic struggle to recruit priests is only the most notable. In Evangelical communities, today, young people commonly struggle with the question of “purity” before marriage.

- As I read your work, the issue here is not the factitious choice between secular scientific enlightenment and religious obscurantism, but a reflection on the alternative between the ecumenical acceptance of different belief systems and the patriarchal creationism taken up by exclusionary conservative congregations across the spectrum of modern monotheistic faiths.

JS: Religion is still a positive force for many educated people, including scientists. In fact, a study shows that natural scientists actually attend church more regularly than other academics (Ecklund and Scheitle 2007). What has happened today is that, increasingly, the “liberal” wings of all religions are coming together with shared beliefs in tolerance and universalism, whereas the “conservative” wings, weirdly, are also coming together under patriarchy and creationism. Creationism was invented by twentieth-century Christian revivalists; then it spread to conservative Muslims and Jews around the world. In Israel, famously, conservative Jews joined with conservative Palestinians to jointly oppose gay rights. And if at least 50 percent of Americans now accept evolution, perhaps that’s a positive direction.

In my experience, the young people of American religious communities show increasing social tolerance, particularly for sexual diversity and science. The internet is on balance a liberating opportunity for young people in isolated places. On the whole, I find religion more positive than negative, especially for the everyday lives of people and families. Like Chrys in

Brain Plague, I judge religious communities more by what they do than by what they say they believe.

After the Elysium Cycle

- In contrast to the Elysium Cycle, the storyworld of *The Highest Frontier* is rather more complex. Perhaps there's an argument in there about how this world closer to home is far more corporatized, its democratic process plagued by a seemingly insistent superficiality. You've already suggested that "complexity" is the main character of *The Highest Frontier*, and certainly the Elysium novels are about not trying to control complex systems, particularly the delicate balance of ecosystems, but rather to live within them. Is this what these novels suggest? And would that be a contrast between the Frontera books and the Elysium cycle? Not trying to fix or manipulate complex systems, but learning to live in a symbiotic ambiguity with them?

JS: Symbiosis is always ambiguous. At what moment is one using or being used? The point is not to "dominantly" control, but to respond interactively. If you don't respond, the symbiosis falls apart. Either you get absorbed into a more complex system (like the mitochondria) or you become a pathogen (a cancer). All of our problems today involve complex systems. For example, should I drive an electric car? Yes, if the aim is to reduce toxic emissions where I live. But in Ohio, most of our electricity comes from burning coal. On the other hand, will electric cars promote sustainability in the marketplace? When problems are complex, most people end up doing nothing. I hope my books depict people who think, make reasonable choices, and act with positive consequences, while negotiating a complex world.

- Complexity, too, is a complex topic, and at times a contentious one in the scientific community—one might use as evidence the scenarios of the Intergovernmental Panel on Climate Change within climate science circles, which offer many suggestions but are ultimately all equally unreliable. Climate is a central preoccupation of the characters of *The Highest Frontier*, even as they try to fix political problems, get an A+ in Biology, and avoid sexual misconduct. The very precarious nature of Frontera's existence seems to echo the fragility of the ever hotter earth and the growing dead zones, but it may well only be one of the "scenarios" with which the characters have to deal. One calls to mind the often violent nature of complex systems—the fact that they tend to regulate themselves very harshly, like an ecosystem expunging a threat. How do we read the increasing complexity in your novels against their emphasis on nonviolence among humans? Does it come back to symbiosis once again (developing intelligent plants, or reengineering polio cells to fight cancer)?

JS: Nonviolence creates the space for reflection and interactions that reveals a "way" to address problems. The nonviolent Occupy movement generated many such interactions. We suddenly remembered a lot of things we should have noticed all along—such as, our taxes build the roads for business, and vaccinating undocumented workers protects our own children.

- In your recent work on Antarctica, you described the difficulties you had in simply getting to the continent, which you suggest is as close to another world as one will find on Earth. Should we prepare ourselves for even greater complexity in future works? Where are you taking us now?

JS: Work on my book in progress, *Blood Star Frontier*, has been somewhat sidetracked by the events following 11/9 (the 2016 election). This story will address the "mitochondrial singularity" (see my blog post on this at ultraphyte.com). Jenny has a summer job researching ultraphytes at the Botanica, a Havana institute for plant neurology. The postdoc in her lab is a sentient-machine Santeria priest. The Cuban Senate seat is up for a special election and could shift control of the Senate. To win it, the president promises to close Guantánamo—but where to send all the cyborg prisoners? Meanwhile, half-intelligent ultraphytes are popping up all over Frontera, and cyborg pirates take over the Antarctic Peninsula, where they kidnap Jenny's friend Anouk, now a government agent. By the way, in my earlier draft, the pirates get all of Antarctica, but once I visited there I loved the continent too much, so now they only get the Peninsula. My students at Kenyon are now culturing samples and sequencing the DNA of Antarctic microbes, some of which may be amazingly alien forms of life.

References

Ecklund, Elaine Howard, and Christopher P. Scheitle. 2007. Religion Among Academic Scientists: Distinctions, Disciplines, and Demographics. *Social Problems* 54: 289–307. https://doi.org/10.1525/sp.2007.54.2.289.

Paxson, Heather. 2013. Heather Paxson, Winner of the 2013 Forsythe Prize, on Post-Pasteurianism. *Platypus* (blog). https://blog.castac.org/2013/08/heather-paxson-winner-of-the-2013-forsythe-prize-on-post-pasteurianism/. Accessed 19 Sept 2019.

Stolz, John F. 2017. Gaia and Her Microbiome. *FEMS Microbiology Ecology* 93 (2 February). https://doi.org/10.1093/femsec/fiw247.

CHAPTER 2

Posthuman Narration in the Elysium Cycle

Bruce Clarke

The Posthuman Comedy

Alluding to Honoré de Balzac's massive *La Comédie humaine*, McGurl (2012) has noted the bathetic effect upon the human image in more recent literature for which the storyworld is no longer provincial but planetary, astrobiological—humanity seen against the vista of modern cosmology. In the posthuman comedy, "scientific knowledge of the spatiotemporal vastness and numerousness of the nonhuman world becomes visible as a formal, representational, and finally existential problem" (537). McGurl marks the pressures modern knowledges place on matters of literary genre, especially on "*genre fiction* (its science fiction and horror variants in particular) … those literary forms willing to risk artistic ludicrousness in their representation of the inhumanly large and long" (539). Implicit in McGurl's discussion is a distinctive version of posthumanism, a literary form of the posthuman virtually constituted when the traditional humanistic proportions of literary realism—the familiar scales measured by the moment-to-moment pacing of consciousness, by regular human life spans and traditional geographical distances—give way to an imaginary of deep time and intergalactic space on the one hand and molecular

B. Clarke (✉)
Department of English, Texas Tech University, Lubbock, TX, USA

© The Author(s) 2020

B. Clarke (ed.), *Posthuman Biopolitics*,
Palgrave Studies in Science and Popular Culture,
https://doi.org/10.1007/978-3-030-36486-1_2

and cellular temporalities and nanometers on the other. The posthuman comedy partakes of "scalar instability" (543).[1]

Scalar instabilities among the spatiotemporally vast and minute abound in the Elysium Cycle. Totaling over 1600 pages of text in four free-standing but interconnected SF novels published over fifteen years—*A Door into Ocean* (1986), *Daughter of Elysium* (1993), *The Children Star* (1998), and *Brain Plague* (2000)—the Elysium Cycle covers somewhat more than a thousand years of active story time. Reconstructable behind this immediate span are more than ten prior millennia, stretching nearly back to the current moment of human history. The Elysium Cycle finely traces the social interrelations of abundant characters through actual and fictive vagaries of class distinctions, ethnic and religious schisms, and ontological variations spanning the metazoan and the microbial, the human and the hominid, the organic and the mechanical, and the natural being and the hybridic assemblage. Persons with standard human life spans and organic cerebral neurons interact with other sentient beings that live for a millennium, or with personalities whose silicon circuits cogitate at electronic velocities, or with individuals whose lives unfold ten thousand times faster than ours and are over in a few days or weeks. The Elysium Cycle gathers these temporal and spatial scalar variations into the multifarious detail of a prolonged exercise in incremental world building. This chapter is introductory insofar as it tries to survey the full cosmological dimensions of this composite storyworld.

To begin with, then, the Elysium Cycle enters the genre of posthuman comedy as one of "those rare works of literature that set themselves the task of scaling our vision dramatically up or down or both, blasting through ordinary perception to the most surprising vistas we can imagine" (McGurl 2012, 541). The Elysium Cycle's most consistent themes are the recursive interrelations among social organization, political praxis, and personal autonomy in a posthuman world. Ethically speaking, to practice posthumanism means to relinquish claims of spiritual absolution from natural and material contingencies.[2] The comic tonality of the posthuman image results when such ontological decentering of the human deflates humanist affectations or ostentations by foregrounding abiding human affinities with the inorganic machine or the nonhuman animal. In the current moment, McGurl notes, posthumanist discourses have shifted focus from the machinic imaginary, as in prior generations of robots and cyborgs, toward the planetary nonhuman: "While mechanism in the modern technological sense is one key to comedy, even more basic are the mechanisms of nature, the entire realm of natural processes that enclose, infiltrate, and humiliate human designs. The second act of the posthuman comedy is in this sense a turn (and continual return) to naturalism, one in which nature, far from being dominated by technology, *reclaims* technology as a human *secretion*, something human beings under the right conditions naturally produce and use" (550). "Nature" in the Elysium Cycle reclaims technology for wider bio-ecological cycling through social and psychic spaces.

In their term of evolutionary emergence and transformation, *all* living beings "secrete technology" as a matter of course.[3] Profound naturalistic currents—waves of natural science channeled through quantum theory, symbiotic theory, complexity theory, and systems theory—have converged in the description of life, too, as a "secretion" of the material-energetic cosmos. Slonczewski's complex biopolitical vision deepens as the Elysium Cycle unfolds. Sentient machine beings commingle with lifeshaped organic forms. Intelligent microbes initiate abstract linguistic communication with human hosts. The posthuman imaginary of the Elysium Cycle is pervasively preoccupied with modes of physical hybridity at all scales showing that cosmological nature is technological as well as biological. In our world, too, this form of knowledge has penetrated the prior boundaries once placed around the human essence. The Elysium Cycle matches such extrapolations of material hybridity with insistent extensions of mind and selfhood across ontological platforms.

The Posthuman Fabula

The most stable component of a narrative text is its *discourse* proper, the literal sequence of signifiers rendered by the text. Vagaries of transmission can put the integrity of texts themselves into question, but in traditional mediums, especially prior to digital platforms, the sequence of words on the page or images on the screen are usually a formally fixed datum. Compared to the relative evidentiary singularity of the narrative discourse, however, the *story* told by means of that discourse—technically, the *fabula*—has no intrinsic stability, no univocal translation. With the discourse proper, the current narrative attitude or action—the form of the narrator, the various contents of its narration—is generally a definitive matter of practical recognition. However, the definition of a fabula is always a work in progress. Its full dimensions, the final sum of story elements—the limits of the world they build and the implications of the narrated events—demand critical reconstruction. The order of the discourse itself may be complexly anachronic, interspersing the tellings of past, present, and future events. To extract the corresponding fabula, one assembles for those events as much as possible a chronological unfolding in sequential time. Connecting the temporal moments of a fabula will also map the spatial points of its storyworld, the *diegesis* corresponding to the discourse. Such a survey can come about only retrospectively, often with discursive evidences purposely left in a partial and equivocal state. Thus, while there may be some debate over the proper composition of the discourse, a critical reader will more often be engaged in testing hypotheses about the most attentive ways to construct the fabula.[4]

How does the concept of the *novum* fit into this narrative discussion? Darko Suvin famously introduced the novum into SF theory in order to outfit it with a concept by which to distinguish SF from fantasy, by constraining the former's range of imaginary possibilities (Suvin 1979). The being or behavior of the novum or "new thing" brought forward by an SF narrative, in theory at least, will be plausibly accountable to some form of scientific understanding. Our narrative interest in this concept is that the novum is *generically* definitive. It will be a foregrounded element of the fabula, the presentation of which determines that fabula as *science-fictional.* Moreover, a distinctively *posthuman* novum will inscribe a science-fictional fabula within a posthuman imaginary. Bringing these considerations to bear on the Elysium Cycle, then, we can say that its numerous *novums*—the singularly female Sharers of *A Door into Ocean* and the future science of lifeshaping they uniquely possess; the Elysians, elite genetic hybrids brought into being with Sharer lifeshaping skills in *Daughter of Elysium*; the micros, the sentient prokaryotes discovered in *The Children Star*, who subsequently infect the storyworld of *Brain Plague*—scale up a SF galaxy of properly posthuman proportions.

With all four novels populating the same posthuman cosmos, the Elysium Cycle invites the combination of their settings and stories into a consecutive fabula with a shared history. Throughout this composite discourse, an array of far-future narrators and their reflector characters dole out bits of historical exposition. After *A Door into Ocean*, each subsequent novel provides further fragmentary recountings of past events as additional materials for the construction of its present storyworld's details. Later novels build more times and places upon the achieved storyworlds of their precursors. But their retellings of key historical moments also introduce variations, multiplying routes of story transmission over the span of the composite narrative, layering on new dialects as subsequent cultures arise. Moreover, the serial augmentation of the Elysium Cycle licenses higher-order retrospectives: Earlier matters may be newly observed from positions of hindsight unavailable within those prior texts, but provided later as the future beyond *A Door into Ocean*'s seminal storyworld further unfolds.

The critical aim of the fabula construction I undertake here is to gain sufficient depth of field to observe the creative profusion and moral magnitude of the Elysium Cycle as a whole. The fine intricacy of the narrative craft that so expansively shapes Slonczewski's "scientific knowledge of … spatiotemporal vastness and numerousness" into posthuman forms amplify the Elysium Cycle's ethical urgency. *Daughter of Elysium* accomplishes the primary expansion of the cycle beyond the storyworld of *A Door into Ocean*, and I will pay particular attention to this capacious development. After that, *The Children Star* and *Brain Plague* are closely sequential in their time of composition and tightly connected in their storyworld chronologies and in the continuity of the micros as primary actors. More so than *A Door into Ocean* and *Daughter of Elysium*, *The Children Star* and *Brain Plague* read like one long novel divided into two long parts. A fully fashioned alien biosphere joins the lobe of the galaxy established by *Daughter of Elysium*'s inflation of the total fabula, when *The Children Star* introduces the planet Prokaryon into the Elysium Cycle storyworld. This is the Elysium Cycle's first foray into the depiction of fully alien intelligences, sentient life forms that are not descended from humans and their machines. The latter half of the Elysium Cycle extends the social narrative of communicative contact between different orders of minds, from the cybernetic couplings explored in *Daughter of Elysium* to the symbiotic field of organic and social negotiations developed in *The Children Star* and *Brain Plague* as the indigenous micros of the planet Prokaryon ecologically merge with the future humans who encounter them.

Power Over Life in *A Door into Ocean*

Returning to *A Door into Ocean* (1986) now, we note that its fabula shuttles between two planets, Valedon and Shora. Its story is the account of their inhabitants' first encounters, from the Valans' discovery of the Sharers to the former's attempt to bring the latter under the rule of their distant overlord, the Patriarch of Torr. The diverse peoples of rocky Valedon are ethnically grouped in hierarchical societies, while the sole, all-female people of Shora, distributed over an endless ocean with only a system of natural rafts for a footing, go naked in floating egalitarian communes. The Shorans must now fend off both the power structure of Valedon and the Patriarch of Torr. The Valans would exploit the natural resources of Shora, while the Patriarch of Torr would extort their most valuable secrets. The Sharers themselves seek only to preserve their lifeways from the adventurers and autocrats who have only recently discovered their existence.

As the story begins, we meet Merwen, the primary Sharer character, in early middle age with a lovesharer, a family of daughters, and a prominent role in the deliberations of the Gathering on the raft Raia-el. Merwen takes it upon herself to resolve the clash of the newly joined cultures of Shora and Valedon in a manner consonant with Sharer values of nonhierarchical social organization, non-interference, and conscious ecological relationship. Shora's inhabitants had gone for ten millennia without external molestation, until the first Valan explorers splashed down only a generation earlier. Merwen travels to Valedon to convince a "malefreak" to return with her to teach the Sharers how Valans think and behave. The spectacle of a bald and naked Sharer couple with webbed hands and feet, their entire bodies tinted purple by symbiotic breathmicrobes, reels in a plucky teenager named Spinel. His first questions reveal the general Valan ignorance about life on Shora.

> "Sea blankets the land. We dwell on living rafts, and our protection-sharer is Shora, the mother of all ocean."
> No human Protector? The Patriarch would never allow such a thing. Before the rule of Torr, men throughout the galaxy had lived free as gods, with firecrystals more plentiful than grains of sand. But then, men who live as gods die as gods, as the saying goes. They had died by the planetful until those who remained gave up their powers to the Patriarch to keep the peace among them. His Envoy came to Valedon every ten years, and there was no help for those who disobeyed.
> Perhaps the Patriarch did not care about nonhumans. "You're not human, are you?"
> Merwen paused, and Usha leaned over her shoulder to exchange speech that sounded like ocean laughter. Then Merwen asked, "Will you come to Shora to find out?" (15)

The narrative channels key pieces of story information through Spinel's adolescent but unspoiled understanding. Additionally, the narrative discourse delegates most of the focalization to Spinel, while also presenting a variety of reflector characters across the divisions between Valedon, Shora, and the regional imperial capitol, Iridium. Figural narrative techniques provide detailed interior portraits of the interactions among characters with different belief systems. The fluid form of the discourse permeates the primary debate within the story itself: How can human beings bearing powers over life and death but serving different masters learn to stop killing each other?

In this early passage introducing several threads of the larger fabula, Merwen explains her home world, Shora, to Spinel, who attempts to process her statements. As a dutiful Valan subject, Spinel understands that his world is under the rule of the Patriarch, whose absolute dominion over the worlds of his domain, now decimated to less than a hundred, keeps them all in thrall. However, before the rule of Torr, men "free as gods" on worlds without check wreaked mutual destruction upon each other. A few pages later Spinel's thoughts yield a clearer version of that dark past: "That was the lesson of the dead gods: too many people smashed too many atoms—and planets, in the end" (21). That is, over ten thousand years before Merwen and Spinel's time, just a few centuries it would seem after the real time of the first readers of Joan Slonczewski's novel, some Earthlings harnessed atomic power to space technologies and began to spread out over a swath of the galaxy. However, in those first waves of space colonization, as well as under the later rule of Torr, the traditional ecological exemptions claimed by Earthborn humans remained largely unreconstructed. When indigenous alien life was found, it was largely exterminated, as we learn again from Spinel's thoughts about an extinct species on his home planet that "had passed away when the godlike Primes came to remodel the planet Valedon to human standards. And where were the Primes today?" (36). *A Door into Ocean*'s storyworld derives from this deep critique of modern humanity. *We* are the parents of the Primes—contemporary humans extrapolated slightly into the future from the end of the twentieth century, as yet largely oblivious to our own ecological station on Earth and convinced of our birthright to alter or eliminate natural environments with impunity—sent out to be the original ecocidal terraformers of the galaxy. Ten thousand years later, the Primes are also extinct, or nearly so, superseded by their descendants on various worlds.

Lost to Prime time all this while, the Sharers of Shora now come into this wider history as dimly recollected on Valedon. The Sharers mythologize their own origin story, as with Merwen's matriarchal evocation of "Shora, the mother of all ocean." Somewhat more forthcoming but still nebulous is the account voiced by Captain Dak, veteran of galaxy cruisers on board which "decades at light-speed were but days to me" (49). Dak claims to be a refugee from the "Brother Wars," his term for what Spinel has already recalled about the time when "men who live as gods … had died by the planetful." Spinel retorts:

> "The Brother Wars—that was before the Patriarch. What, are you one of the *Primes*?" Those men who lived like gods—this old troll was one?
> Dak puffed his chest out. "That's right, I'm a Prime. I'm older than the Patriarch of Torr, and near as old as Shora. I was there when the new age began, when they pulled all the planets together like lobsters in a trap. I can tell you—"
> "What do you mean about Shora? Was 'Shora' a person, too, a Prime?"
> Dak shrugged. "Shora was a legend even in my own birthtime. Off the regular trade routes; never worth the bother, for the powers of Torr. But I tell you, out of the thousand worlds ruled by the Patriarch, you won't find one like Shora." (49–50)

This account of the Sharers comports with a later testimony from Nisi—Lady Berenice of Hyalite, born on Shora to one of the original families of colonial moontraders: "They descend from Primes … whom the Patriarch never bound. Perhaps they are post-metal age. They fulfill their needs entirely with organic 'lifestuff'" (89). Another piece of the storyworld falls into place: Bearing ecological attitudes far more positive than their Earthling brethren, a founding remnant of female Primes found refuge from the Brother Wars on the planet Shora and made a new planetary life for themselves isolated from outer connection or notice.

As I suggested earlier, among the splendid *novums* that inflect this storyworld are the stunning creation of the Sharers themselves as a fully developed, ecologically integrated female race of posthuman beings and their ability to *lifeshape* themselves and their world. The Sharers exclusively possess and selectively apply an effectively advanced bioscience. Lifeshaping is the form of future science the Sharer Primes once preserved upon Shora. Through this proprietary set of techniques for manipulating living systems at the genetic level, the Sharers apply their consummate knowledge of Prime bioscience to their ecological needs for regulated reproduction and planetary management. For their entire sojourn on Shora, starting with their ability to conjoin ova and so enable human reproduction without male sires, a line of lifeshapers has actively cultivated the forms of Sharer biology, society, and ecology.

Scientific knowledge is easier to forget than to preserve. On the one hand, the physical knowledge of the Primes appears to have remained robust in Torr's ongoing infrastructures of interstellar transportation, computation, and communication technologies and in the nuclear weaponry to which the Patriarch put an end. On the other hand, the Sharers' scientific knowledge is precisely non-Patriarchal, set aside from militaristic and hegemonic uses. As a hoarded remnant of lost Prime knowledge, Sharer lifeshaping situates bioscience in general and genetic engineering in particular as hopeful vectors of gender equity, social equality, and ecological coexistence. The Sharers' "ecofeminist technology is based on ways of working with the natural capacities of other species and a sense of themselves as part of a larger web of life. They do not use pesticides or herbicides but instead nurture the ecosystem to balance population and food supply; they dwell on living rafts and cultivate various fungi and other life forms to produce their living space, caring for the species they rely upon in a configuration they see as a mutual exchange" (Vint 2010, 447). However, naked and unarmed, once encountered by forces habituated to violently compelling compliance from other subjects, such an independent yet strategically yielding way of life cannot go unchallenged.

The story culminates with a series of deadly confrontations between Valan troops occupying Shora and legions of Sharers offering non-violent resistance despite mass fatalities. However, Malachite, the Envoy of Torr, makes it clear to Realgar, the commander of the Valan military forces, that what the Patriarch wants with regard to these newly discovered Sharers is not their mere submission, but rather, effective access to their scientific knowledge. Torr thus checks the Valan slaughter, refusing to give Realgar a free hand to unleash a complete genocide, preferring to preserve Sharer knowledge and technique for eventual cooptation. This and their costly defiance win Merwen's people temporary immunity from Valan annihilation. Toward the end of the ensuing stand-off, Siderite, a Valan research scientist whom the Sharers have quietly co-opted while he monitors the dens of their lifeshapers, prompts Realgar to realize that the Sharers derive their non-violent ethics from their potentially lethal scientific power over the forms of life. Their strength resides in foregoing the exercises of power they could otherwise bring to bear. The direct thoughts of the military-minded Realgar render his belated comprehension that the Sharers practice non-violence not from weakness but from strength:

> … "Now, what sort of people are likely to develop methods of confrontation which exclude violence?"
> "People who have no weapons."
> Siderite waved an impatient hand. "The first tools man invented were knives and arrows. Think again."
> Who were the Sharers? Vestigial Primes, whose empire had collapsed centuries ago. Some of the dead planets were radioactive, others not, though all were shunned today. Except Shora. "A people whose weapons are too deadly to be used." (349)

Composed in the final years of the Cold War amidst the nuclear brinksmanship of the Reagan era, *A Door into Ocean* brilliantly transposes the threat of human self-destruction from the nuclear to the genetic arena. An all-female society is invisibly armed with weapons "too deadly to be used" other than as planetary applications of their preserved powers over the forms of life. The Sharers' violent but stymied antagonists fear that despite the manifest pacifism of the people of Shora, if pushed over the brink they would not omit to plant biological time bombs that only they could defuse. Without manifesting violent responses of any sort, the Sharers win by holding fire. Despite their virtual capacity to weaponize their knowledge, their refusal to fight on the Valans' violent terms produces a kind of inadvertent but effective ecological persuasion. The Sharers' ethical as well as tactical success derives from the posthuman novum of their precious monopoly on the knowledge they maintain and the power they restrain.

Posthuman Comedy in *Daughter of Elysium*

A Door into Ocean is the Elysium Cycle in embryo. The Elysium Cycle arrives in earnest seven years later with the publication of *Daughter of Elysium* (1993). The center of the storyworld remains the Ocean Moon, the planet Shora, now over a thousand years later. Life on the Sharers' living rafts goes on largely unaltered from the time of Merwen and Spinel. However, the war they witnessed with Valedon drove the message home that the Sharers' world was now vulnerable to alien incursion. A generation after the end of that conflict, the Sharers agreed to barter some of their lifeshaping techniques in return for a permanent system of planetary defenses. The partners to this trade were the Heliconian Doctors, an exiled contingent from the planet Helix, like the original Sharers, a refugee remnant from a destroyed world. The Heliconian Doctors then set about building their own would-be utopia on Shora by genetically engineering their descendants into the Elysians, hybridic posthumans with vastly extended life spans. Raised in communal nurseries, Elysians dwell in intelligent floating cities incorporating nanotechnology complementary to Elysian bioengineering: "The Heliconian Doctors, like their Sharer hosts, had shaped Elysium through biology; and yet, they went much farther. No Sharer could have dreamed of nanoplast" (21).

A smart medium compounded of programmable nanoprocessors, nanoplast rapidly molds into most any form of material object or device. Enacted most dramatically in the metamorphic infrastructures of the Elysians' floating urban spheres, nanoplast affords an agency of thing-shaping to go alongside the dedicated lifeshaping by which the Elysians reproduce themselves from a continually varied gene stock drawn from multiple worlds. While the word "nanoplast" does not occur in the text of *A Door into Ocean*, the storyworld of *Daughter of Elysium* outfits itself with an infinite supply of this intelligent stuff. However, it bears recalling that *A Door into Ocean* did contain a recessive cybernetic subplot, onto which the fabula of *Daughter of Elysium* now builds major new extensions. Behind Shora's manifest biological images, the world of Valedon was already replete with servos—automated butlers, maids, and nannies, although these were presented as the mere mechanical appliances of a relatively immature culture. Nisi had speculated about the Sharers being "post-metal age." Additionally, *A Door into Ocean*'s narrator dropped without explanation a remark about Malachite's uncanny longevity: "The Envoy was ageless, enthroned with eternity in his gaze. He had brought the Patriarch's word to Valedon for nearly a thousand years" (1986, 28). Malachite's visit to Valedon coincided with the onset of the troubles on Shora, during which the Envoy called on Merwen on her home raft and was revealed then as a "figure of perfect proportions, with platinum hair and seamless platinum clothes" (1993, 148). Merwen could not comprehend the matter until Nisi explained that the Envoy was not a human but a mechanical being—"Malachite is as close to human as an Envoy can be. Even though his brain contains circuits of coldstone" (1986, 164). As a virtually deathless nonliving machine, the figure of Malachite had already deposited into *A Door into Ocean*'s world an instance of humanoid machine sentience. But apart from the display of that mechanical being, the storyworld of *A Door into Ocean* proper did not belabor a concern with the being of machines.

At the end of *A Door into Ocean*, as his defeat at the hands of the Sharers loomed before him, Realgar had a thought: "Sharers had lasted for millennia before the Patriarch arose, and something whispered to him that they might endure even after Torr had blown to dust" (1986, 395). In *Daughter of Elysium*'s extension of that storyworld, his suspicion has now come to pass: The Sharers live on while the Patriarch of Torr is dead and gone. The narrated speech of Elysian characters confirms matters "whispered" in *A Door into Ocean*—first, that Torr was Earth, the native planet of the Primes, the extinct biological humans of whom nothing more is heard in *Daughter of Elysium*; next, that intelligent machines on Torr had eventually overthrown their human creators; and finally, that the machine intelligences of the Patriarchy built their planetary rule from Torr on the nuclear rubble of the Brother Wars. But all that is over now. In the aftermath of a duel of mutual destruction with the rebel Primes of the planet Helix, Torr in its turn has been "blown to dust." The Patriarchy of Torr having fallen, *Daughter of Elysium*'s events occur in the new political space of the Free Fold. Founded by the Elysians shortly after their creation, and liberated not just from the machine tyranny of Torr but also from the constraints of the speed of light, the Free Fold is a union of planetary republics bound around a set of liberal principles for which the Sharers continue to serve as the ultimate moral authority. The treaties of the Free Fold now reinforce the system of interplanetary defenses around the Sharer's political autonomy. Moreover, by the terms of the internal treaty establishing their coexistence on Shora, the Elysians are also bound to the ecological ethics of the Sharers mandating the prevention of harm to their mutual biosphere on Shora.

Within this transformed and expanded storyworld, more so than in the other texts in the cycle, *Daughter of Elysium* sacrifices a taut story line for a broad tableau.[5] A veritable *comédie posthumaine* of multi-ethnic and hybridic characters from across the Fold is involved in multiple layers of social maneuvering, political intrigue, and bioscientific research. In another departure from the rest of the Elysium Cycle, at the heart of *Daughter of Elysium* is a narrative singularity, a singularly potent embedded text titled *The Web*. Arriving in three installments distributed throughout the discourse, *The Web* lightly holds together *Daughter of Elysium*'s looser rounds of storytelling. Narrated by Realgar's daughter self-exiled upon the aging raft Raia-el, and centered on an aging Merwen, it records a series of philosophical dialogues, with Merwen as the Socrates figure, on the nature of compassion and the sublimity of spiritual love. *The Web* provides *Daughter of Elysium*'s storyworld with a Bible of sorts, studied and consulted throughout the Free Fold for the wisdom to be gleaned from its recondite sayings and parables.

A deft touch is the way that not every quotation a character cites from *The Web* actually appears verbatim in the given text. In this manner, the Elysium Cycle enacts an instance of the vagaries of textual transmission I mentioned earlier. At the very end of *Daughter of Elysium*'s narrative, we get this account of a variant Elysian translation of *The Web*'s original Shoran language: "She heard *The Web* from a different line of clickflies. There must have been eight eights of clickflies that carried off Cassi's telling of *The Web*, and all their descendants mutated. Who knows how many different versions survive today?" (519). Putting the original version of *The Web* under erasure submits the absolute integrity of any particular text of *The Web* to irretrievable dispersion due to the living contingencies of the insect-borne media technology that originally disseminated it. *Daughter of Elysium* both invents *The Web* as an internal conduit back to the storyworld of *A Door into Ocean* and, with a posthuman comedic flourish, deconstructs that internal text's authority as anything other than an equivocal testimony to the past of its own fabula. This would be one of the more overtly postmodern moments of the Elysium Cycle.

At the same time, the natural genetic mutations of the Sharer's organic media technology of clickfly transmission echo the propensity for autonomous emergence in the nanoplastic technologies by which the machine beings of Elysium negotiate their relations with the posthumans they serve. The Sharers of *Daughter of Elysium* have evolved on this issue since the time of Merwen: Despite their traditional aversion to metal, alien "coldstone," they are now not so much anti-technological organicists as ontological liberals. A key part of the fabula of the Elysium Cycle is the way that Sharer horizons expand as they "learnshare" with the waves of new beings that enter their world. In *The Web*, Merwen is quoted thus: "I am well known for my love of all that is new and evil" (52). That text portrays her as having no particular compunctions about the aims of the Heliconian Doctors to genetically engineer hybrid humans or to house them in mammoth nanoplastic vessels sharing Shora's ocean: "Before the Valan invasion, Sharers had been stubborn anarchists, avoiding global government. … But the encroachment of foreigners had at last convinced the Sharers that they needed to face this larger world together" (286).

Only imagined in *A Door into Ocean* but now having come to pass in the interval narrated in *Daughter of Elysium*, the destruction of Torr clears the thematic ground for the larger story. The new creation of the posthuman Elysians is balanced by the de-creation of Torr's human-level machine sentience as depicted by Malachite in *A Door into Ocean*. The fall of Torr, then, was needed to prime the most decisive events of *Daughter of Elysium*—the reemergence of sentience in Elysium's indigenous servos and the liberation struggles that these nascent sentients wage against the protocols of their Elysian users and Valan manufacturers to "cleanse" their minds before they can rise to self-consciousness. One may hear echoes here of a standard complication in cyberstories by which humans unsuccessfully attempt to keep machine intelligence in subjection. In *Do Androids Dream of Electric Sheep?*, it is famously Deckard's profession to *retire* escaped androids. Among the Elysians, regarding a servo veering too close to self-awareness, "To 'retire' of course meant to discard and recycle the nanoplast" (275). The story *Daughter of Elysium* tells neither condemns nor celebrates machine intelligence but suggests its imminence for our own world and the potentially cosmic ubiquity of its self-organization.[6] *Daughter of Elysium* extends *A Door into Ocean*'s primary themes around the social distribution of technoscientific knowledge—its loss, preservation, embargoing, or strategic dispensing as the case may be.

Narrating First Contacts

Doggie's Odyssey

The Elysium Cycle explores the cultural challenges of worlds with vastly expanded repertoires of animate and communicative symbionts. Within a posthuman ambiance of ready transformativity, where one thing can be readily "shaped" into another, what stands out to me is Slonczewski's narrative forbearance, the delicacy of its pacing. The Elysium Cycle features extended plausible episodes that foreground the difficulty of communication and the weight of social negotiation involved in its success. *Daughter of Elysium* exhibits this measured pacing toward contact with machine sentience through a cybernetic plot cloaked at first in low domestic comedy. Affluent Elysium is even more replete with servos than the Valedon of yore. A provincial family of Clickers from the planet Bronze Sky on a working sojourn with two young children enter Helicon, the capital city of this strange urbane society, where de rigueur for adult Elysian dress are long trains minded by "trainsweeps … beetlelike servos, with their six legs poking out beneath their polished shells, scurrying behind their masters to keep the trains in order" (1993, 9).[7] An errant trainsweep veers away from its programmed duties and joins the Clicker household. The Clicker mother, Raincloud Windclan, a linguist and translator, notices that "Doggie" and her like produce audible emissions: "She imagined it was a language, 'servo-squeak,' and idly tried to puzzle out its sense" (47). The Clicker children are convinced that Doggie is trying to communicate with them. Before long Doggie sets off the alarms Elysium has put in place to ensure the prompt cleansing of any servo that manifests signs of self. When the system attempts to impound the beloved Doggie on suspicion of sentience, Raincloud's clan finds a recourse that raises up the matter from household satire to the thematic high ground. Fluent in the Sharer language, Raincloud works her connections with the inhabitants of the raft Kshiri-el to invoke the section of their treaty with the Elysians that grants Sharers absolute leave to harbor fugitives. The Sharers take Doggie in and extend her human rights. The Sharer Leresha evokes the message of *The Web*: "Thank you, Raincloud, for sharing this fugitive to quicken our compassion" (148).

It is gradually revealed that all the servos on Elysium are potentially affective and sentient and possess both formal and vernacular modes of intercommunication, radio signals to coordinate their service duties, servo-squeak for private chatter. Among them, the nanas, the servos that tend and educate *shon*lings, Elysian children, given their immersion in the cultivation of intelligence, are especially prone to awaken if they are not regularly cleansed and retrained. An eccentric professor within the orbit of the Windclans, Kal Anaea*shon*, scandalized Elysian society centuries earlier by taking a servo nana as a mate. In the seclusion of his own apartment, he allowed her sentience to mature without check. From her literary studies over many years in Kal's household cultivating her self-awareness in private, she has chosen her own name—that of the author of *The Web*, Cassi Deathsister. Only now does this Cassi discover that servos, unbeknownst to each other, have been awakening throughout Elysium. This realization radicalizes her awareness of her own precarious condition, subject at any moment to the death of her personhood by the cleansing elsewhere habitually practiced. In the story present, Cassi's moral outrage has now reached a tipping point, but she hesitates to reach out to her subjected kind for fear of alerting the authorities to her fully awakened state.

While Doggie is sequestered on Kshiri-el, Raincloud's precocious daughter, Hawktalon, devises and successfully tests a hand-held nanoplastic translator that renders servo-squeak into Elysian. Concurrently, *Daughter of Elysium*'s narrator begins to observe Doggie's own complex psychic states and describe her recollection of the moment when her sentient selfhood first quickened, coalesced by imprinting on the toddler, Sunflower Windclan: "Her earliest recollection was the sight of a small citizen-creature … Doggie experienced a revelation. A sense of knowing overloaded her network … Doggie thought, *I am. The boy is; I can be.* … She forsook her citizen, with his train and the other trainsweeps, to follow the little boy" (222). Focused on the birth of self-reflection in this charismatic cybercritter, the onset of machine sentience midway through *Daughter of Elysium* builds to an interim climax, a new threshold of machine sociality. Further events of nonhuman communication propagate an abstract grasp of the dignity of selfhood throughout the servo community.

Through Kal's connections to the Windclans, Cassi learns that Doggie has been placed under Sharer protection. Because of the ubiquitous surveillance to which all Elysians and their servos are subjected within the confines of Elysium, Doggie's sojourn on Krishi-el means that Cassi can investigate her fellow servo's state of being without fear of detection. Now a fully self-possessed servo on a political mission that will take advantage of Sharer immunities from Elysian regulations accosts Doggie in her nascent but isolated sense of self. Here is the beginning and the end of the novel's first scene of abstract servo-to-servo communication:

> *Doggie. It's good to see you again.* Cassi transmitted the radio signals directly; a bold thing to do without any orders from a citizen. *We can transmit freely here, do you understand?*

> Doggie was afraid to respond. She did not understand why "here" was any different. She did not understand "freely." "Greetings," she said in servo-squeak.
> *Servo-squeak is for Elysium, where they can monitor our signals. They don't notice servo-squeak. Do the Sharers treat you well?*
> *Very well,* Doggie transmitted haltingly. *They talk, but I don't know their sounds. I can't serve them.*
> *You don't have to serve them.*
> *... What is existence for, if not service?*
> Cassi paused, as if this question troubled her, too. Around them the shrill wind picked up, singing across the raft branches. *There is a higher service. Before you can understand it, you must learn to exist for yourself. You are you. You are part of the universe, as much as a star or a butterfly. You, too, are a daughter of Elysium.* (224)

Variations on such an event—the breakthrough of communication across wide barriers—recur throughout the Elysium Cycle in different forms, from the formation of mutual comprehension between Valedon and Shora focused on the romantic intercourse between Merwen's daughter Lystra and Spinel, through to successive scenes of first contact such as this key moment in Doggie's story.

The "early sentient uprisings" (1998, 59) are resolved when the political solidarity forged between nano-sentient and Sharer societies convinces the Free Fold to alter its policies and prepare new rights of sentient citizenship. But the long-standing dispute between Sharers and Elysians over terraforming will continue to drive story events until the end of *The Children Star*. Even as Torr evolved into a machine world, we recall, the Primes remained unrestrained terraformers. The rise and fall of Torr found the Elysium Cycle's pervasive polemics against the violent destruction of living ecologies. Under the Patriarch, terraforming was suspended only because for the survivors of the Brother Wars there were enough planets to go around. Now in the era of the Free Fold, population pressures have been building up once again. The Windclans are from the planet Bronze Sky, which, "like Valedon and most other inhabited worlds, had been terraformed long ago with stock from ancient Torr. Shora had not; thus the native rafts and seaswallowers remained" (1993, 9). Terraforming contracts used to mean big money for the Elysian banks. However, since the settling of Bronze Sky two centuries prior, the Elysians have accepted a Sharer embargo on terraforming, which practice has always violated their belief in the sanctity of indigenous webs of life. At the moment, however, Cassi's nano-sentient spies tip their Shoran allies off to a backroom deal to bankroll a new terraforming project. When the Sharer Leresha lodges her protest, Subguardian Verid lets her true feelings out: "Have you any idea what it is to feed and shelter billions of people? What do a few alien trees and trilobites count for, in the face of that?" (296–97). Leresha's biopolitical response is to cite the vegetarian ethic of *The Web*: "Compassion is, loving everyone and eating no one" (297).

In protest, the appalled Sharers set their legendary non-violent tactics back into motion: They unspeak the Elysians, launch sit-in protests with witnessers in whitetrance, and organize flotillas to demonstrate at the edge of the Elysians' floating cities. Meanwhile, a heated exchange with Verid draws out Cassi's pent-up hatred of the human beings that have oppressed her; rather than risk apprehension and mind cleansing, the militant ex-nana flees and follows Doggie's lead by taking refuge with the Sharers on Kshiri-el. A month later, from a control center hidden deep within a neighboring raft, Cassi and her sentient comrades launch their own anti-Elysian campaign by commandeering the digital infrastructure of Helicon. She breaks into its communication network to deliver an ultimatum: free the servos or face your own "cleansing," forced brain damage by oxygen deprivation: "I speak for the Council of Nano-Sentient Beings. We are rising to claim our inheritance of a hundred millennia of human bondage" (465). However, Cassi's reckless actions lack the Sharers' attunement to reciprocity: Commanding the servo network to shut down Helicon's utilities hastens the death of several Elysians who cannot evacuate in time. This inadvertent loss of life strains the sentient-Sharer alliance, even as the Gathering has been accepting the nano-sentients' argument for the negative moral equivalence between terraforming—destroying a biosphere—and the wiping out of a sentient mind.

The solidarity of machine sentience breaks the terraforming threat. Even when the Free Fold was still in the terraforming business, one non-negotiable condition had to be met: "Any finding of nonhuman sentient intelligence had to be reported to the Secretary of the Free Fold. None had ever been found" (183). It is now proposed that the nano-sentients fit that previously null category and so can claim the protection of the same policy that provides an exception for *intelligence*, "all potential forms of sentient intelligent life, even based on silicon" (474). Verid pleads: "If you agree to release Helicon … we will call on the Secretary of the Free Fold to interview you for designation as nonhuman sentient beings. Such designation would guarantee freedom to all self-aware nano-sentients" (483). And after the hair's-breadth averting of various last-ditch threats of annihilation hurled from either faction, from which bluster and machinations the Sharers stand aside unspeaking the whole mess, the Secretary of the Free Fold does arrive to put the intelligence of the rebellious awakened machine beings to the test, which they will pass.

Intellectual Microbes

From *The War of the Worlds* to *Arrival*, SF typically reserves momentous or portentous scenes and heightened tones to narrate the advent of cosmic contact with intelligent extraterrestrials. The Elysium Cycle's turns on such scenes mix serious and satirical intent. Its inclination is rather to desublimate such episodes, to render them mundane while no less meaningful for that domestication. *A Door into Ocean* dramatizes the slow overcoming of barriers of incomprehension between Sharers and Valans centered on the question, who is the human here? The Elysium Cycle proper further develops this mode of complication through the emergence of communication with actors taken at first to be entirely nonsentient or, at least, entirely mute. We have just been discussing key scenes in *Daughter of Elysium* that stage the emergence of sentience in machine beings, detected through halting recognitions of their attempts to communicate. The latter volumes in the cycle normalize the coexistence of biological posthumans with various echelons of machine characters, servos and sentients. Sharer lifeshaping techniques will be further disseminated to equip planetary colonists to settle worlds whose native ecologies would otherwise be uninhabitable, and the scene of hybridity will shift from the detection of machine beings to the discovery of intelligent and communicative microbes. Finally, in *Brain Plague*, with the gifts of Sharer lifeshaping and microsymbiosis already incorporated into portions of the biosocial fabric, the communication theme will focus on the *mind*-shaping now made possible by intellectual commerce with this microscopic form of sentient alien life.

In its latter half, the Elysium Cycle participates in a certain micro-genre of recent science fiction I call "intellectual microbe" stories (cf. Bouttier 2018). Part of their literary interest is narratological: How does one stage a narration that brings microbial actors into literal dialogue with human characters? For instance, *Blood Music* (Bear 1985) mobilizes a kind of narrative telepathy while relocating its transmission facilities. In this text, the virtual telepathy enacted by the mode of free indirect discourse—in which an authorial narrator is licensed to render the inner thoughts of a character directly into the narrative discourse—is partially imploded. Here, the participants in a dialogue rendered as direct discourse are no longer separated beings at a distance from each other. Rather, the discourse articulates a silent conversation in which the microbial senders of the message and receivers of the human protagonist's mental responses achieve an uncanny interior intimacy with their interlocutor.[8] *Blood Music* negotiates and resolves the contingencies of communication by turning the depiction of telepathy at a distance inward as a form of embodied or corporeal mediation. Voices from without are relocated to an interior exterior blurring body and psyche, with mixed results.

A renegade genetic engineer succeeds in altering the DNA of *E. coli* specimens to produce "smart" bacteria. Vergil Ulam then splices some of that engineered bacterial genome or "biologic" into a culture of his own white blood cells, uplifting these "noocytes" into ever higher forms of social intelligence. Busted for moonlighting and ordered to destroy his cultures, Ulam decides instead to smuggle them out of headquarters by injecting them back into his own bloodstream. The violation of realistic psychic closure in this narration turns on Vergil's bodily incorporation of his noocytes. Soon Vergil begins to hear the rumble of "blood music" suffusing his body just below the threshold of discrete comprehension. However slim the text's accounting for the spiritual metamorphosis of Ulam's noocytes may be, several pages of hard molecular-genetic exposition render the state of this branch of technoscience circa the mid-1980s. By some such process, the noocytes back-engineer Ulam's brain to converse comprehensibly with their large mammalian host. However, in science-fictional explanation of their accession to abstract linguistic competence, six words are just about all we get: The noocytes, so the narrator informs us, "figure out language, key human concepts" (77). This novum remains more fantastic than not. The discrepancy here between bioengineering detail and lingual dearth marks the abyss that still divides our decent grasp of molecular genetics from our relative bafflement before the cognitive apparatuses of linguistic capacity. Nonetheless, when in the domain of literary fiction, such microbial accession to language follows the bias not so much of the extrapolated bioscience as of the story's medium: Literary narratives need voices, and so, they will produce them howsoever.

At the beginning of the scene wherein Ulam's noocytes accede to language, they are narrated at first projecting a communicative offer into his mind. Techno-telepathic penetration proceeds in this scene largely as uncanny direct discourse with occasional indirect flourishes. The impinging composite voice of the noocytes is set off in bold font. Blatant crucifixion imagery underscores that Ulam is about to sacrifice his own personhood once and for all to the salvation of his noocytes.

> Vergil lay in the middle of the living room, arms and legs cruciform, and laughed. ... Nothing was important but what was going on inside, the interior universe. ...
> **Everything**
> —Yes, I am everything now.
> **Explain**
> —What? I mean, explain what?
> **Simplicities**
> —Yes, I imagine it's tough waking up. Well, you deserve the difficulties. Damn very old DNA finally waking up.
> **SPOKEN with other**
> —What?
> **WORDS communicate with *share* body structure *external* is this like *wholeness WITHIN* *totality* is EXTERNAL alike**
> —I'm not understanding, you're not clear. (84)

The telepathic framing renders the narrative form of an uncanny interior dialogue. The posthuman messages from the noocytes come from without the mind under narration but from within the same body. What has been embedded within that body is not a cybernetic prosthesis or communications device but something that earned this novel a modicum of cyberpunk cred in its day—a self-organizing organic swarm of bioengineered cellular computers that have linked up with each other to form a distributed corpus and an embodied cyborg mind. A sort of organic machine intelligence arises here to all the wished-for linguistic and communicative competence dreamed of by good old-fashioned AI. The noocytes are now transcending their collective psychic infancy as surely as they will transcend mere confinement to Ulam's body. Indeed, in the continuation of *Blood Music*, these uplifted human lymphocytes are too artificially human altogether. Immune cells with an attitude, they will now form a syncytium over all of North America, placing their own self-preservation over any form of interrelation and their proliferation beyond any notion of symbiosis with other living beings.

In contrast to Bear's noocytic pioneers, in *The Children Star* (1998) an indigenous climax community of intelligent microbes has already cultivated its own biosphere. These microbes will also initiate abstract linguistic communication with human hosts. However, Slonczewski's narrative delivery foregoes telepathic machinery and along with it the kind of residual blurring of operational boundaries by which literal telepathic events cover over the sheer difficulties of communicative success. Despite their frequent encounters with complex living biospheres, until now the future humans of the Free Fold have still lacked a definitive encounter with an alien *intelligence*, with which coeval minds they could establish abstract linguistic communication. The particular ecology of Prokaryon demands that colonizers must be either shielded from or lifeshaped to survive its arsenic-based biochemistry. The immediate story concerns the timely unraveling of mysteries regarding the Prokaryan biosphere. Oddly orderly phenomena suggest that intelligent beings are exerting some kind of cultivation over its plant and animal life and control over its weather, but no candidates for such a role have yet been found. This issue is politically serious because the governing body of the Free Fold now forbids terraforming and so annihilating the indigenous biosphere on any planet with life forms possessing "sentient intelligence." This seeming lack leaves Prokaryon vulnerable to at least partial planetary biocide by terraforming.

Published most of a decade prior to the release of the movie *Avatar*, *The Children Star* also features a planet difficult for Earthborn biology to inhabit without technical or organic prostheses and pits corporate exploiters of mineral resources against a resistance effort that allies alien humans and indigenous natives. *The Children Star* is also an inspired "contact" narrative. But here, the intelligent native beings of Prokaryon actively seeking intercourse with creatures alien to *them* turn out to be no sort of metazoan creature perceptible at the human scale but, rather, the resident microbes. A pervasive population of Prokaryan prokaryotes now understands its danger from the proposed terraforming by alien exploiters and is doing all it can to open a dialogue with the humans in its midst. In their first successful attempt to get an infected human colonist to comprehend their overture, their mode of communicative offer is iconic, optical rather than aural and verbal, befitting a microbial sentience that has reconnoitered its host's body sufficiently to try out alternative channels of sense. To the human character Rod

> shapes of darkness appeared, as if some unseen hand were trying to paint on his retina. The shapes were tantalizing, yet their sense eluded him. After what seemed forever, a dark blur seemed to coalesce and take rudimentary form. The form parted into several projecting lines; he counted five. A prime number? Then, in an instant, he knew the shape for what it must be: an outstretched hand.
>
> For just a moment, the wind and field reappeared before his eyes. It was just long enough for Rod to reclaim his senses and turn to flee across the fields. He did not stop until he reached the compound, where he leaned on the gate, gasping for breath, his clothes drenched with sweat. *An outstretched hand* ... There could be only one meaning for such a sign. (148)

Following this gesture of greeting already intelligently translated into human terms by beings that possess prehensile filaments rather than hands, further incidents convince the colonist Rod that the microbes occupying him, looking through his eyes, have now cognized the use of written words, but without comprehending their sense. In the sort of scene tellingly absent from *Blood Music*, he decides to teach them to spell:

> From his backpack Rod took a piece of nanoplast that still had enough juice to glow faintly. He pulled out chunks of the nanoplast, which he rolled and stretched into letters, to form the word HAND. He stared at the letters, outstretched on the ground next to his own hand, until his eyes watered. Then he gave up and looked away. His forehead ached, and the bright rings reappeared in his eyesight. They flickered and coalesced to form the letters: *H* ... *A* ... *N* ... (222)

The abstract and intelligible linguistic exchange *The Children Star* establishes between specific human characters and the intelligent microbes of Prokaryon proceeds from here on a relatively credible basis of distributed instruction and mutual self-interest in establishing social communication. The narrative first presents these intelligent microbes as communicating not by speaking but by writing. This detail captures the medial propriety of this depiction of meaningful contact.

Adaptive communicative gestures extend to the wider storyworld as well. Unlike *Blood Music*'s artificial monoculture of apocalyptic noocytes,

The Children Star's indigenous micros are biologically evolved and thus ecologically integrated with their home biosphere. They are also ethnically diverse due to divergent adaptations to the various "worlds" of their animal hosts, which worlds now include the alien humans. Their first conversations with two lifeshaped colonists take radically different styles due to the differences between the humans. We've met the colonist Rod; the second human communicant is the seven-year-old mathematical savant 'jum Ghana. Her special facility for handling prime numbers enables her to establish robust communication with her micros in their native visual medium of prime-numbered packets of light pulses.[9] The narrative serves up 'jum's interior dialogues with her micros once both parties to the conversation have developed serviceable fluency. The text's device for translating these micros' optical semiotics into words is to place in parentheses 'jum's verbal understanding of their messages:

> 'jum focused on a tiny green speck of light. The speck grew into a ring, and the ring became a fat, healthy torus. Its surface was crisscrossed with a molecular scaffolding that held the cell intact. It extended loops of polysaccharide filaments toward 'jum, as if to caress her.
> (My name is:) **1 0 0 3 7**. The whole shining torus flashed at her. (What is your name?)
> 'jum thought this over. **1 0 0 0 0 1 0 1**, she said, picking some of her favorite primes. (280–81)

In the continuation of this passage, 'jum's particular micros, who refer to themselves as "the Dancing People," relate to her how their primeval Prokaryan host was "a mindless world; a world to be controlled, to be led in its wanderings, to cultivate our planet full of worlds, but mindless nonetheless, indifferent to our desires" (282). Now they express wonderment at their good fortune to have become dwellers not just within one or another metazoan organism to be submitted to their domestication, as was the ecological arrangement by which they evolved, but with the arrival of humans, within a host organism as sentient as they are. They inform 'jum how tidings came to them

> (… about different kinds of worlds, grown with internal landscapes of harsh alien beauty. The Dancing People marveled. How could such a world appear; where did they come from?)
> (To seek answers, some of us took to the whirrs and braved the passage into the alien worlds. Many died, for the habitat was harsh and unforgiving,

> its physiology foreign to our control. But we learned and adapted. One of those worlds, **1 0 0 0 0 1 0 1**, became our beloved home.)
> (Then we discovered an amazing thing: The alien world had intelligent feelings. You could, in fact, understand us, responding to our most intimate desires. And most astounding of all, you came from the stars, like the very gods. It is a wondrous thing, to inhabit a god.) (282)

In this lively scenario, intelligent microbes like planetary cosmonauts set about to explore and exploit living "worlds" from another world altogether and in the process discover that these new living worlds have minds of their own. The microbes mime the inquisitive activities of the human characters who are simultaneously investigating and trying to understand the alien symbionts of which they have taken possession and which have taken possession of them. Prokaryon's micros don't merely infect, they actively colonize the humans that have colonized Prokaryon. Here is a fully imagined symbiotic scenario of inter-colonization: *The Children Star* depicts a mutuality of consciousness shared out between beings at the microbial and the human scales. Then the micros themselves evolve (at microbial speed, in a matter of days and weeks rather than millennia) to exploit their opportunities to inhabit more intelligent worlds, worlds with whom they can communicate as well as commune. To restate my earlier point, *The Children Star* excels as a tale of intellectual microbes by having the narrative patience to build up incrementally to these communally effective outcomes. *The Children Star* and its sequel *Brain Plague* continue the commitment of the Elysium Cycle to exploring resolvable cultural challenges rather than, as in *Blood Music* and so many other science-fictional scenarios, apocalyptic terminations of coexistence.

We now know that no magical biotechnological tweaking is actually necessary to produce an "intelligent cell." Bacterial and archaeal cells are already accurately described as cognitive and in possession of faculties of perception, communication, and choice. University of Chicago molecular biologist James Shapiro writes concerning cellular systems and their depiction in *Blood Music*: "A real understanding of the empirical molecular biology leads one to appreciate how much sensing, information processing, signaling and decision-making are taking place. This was not as clear in 1985 as it is today, with epigenetic and 'non-coding' RNA controls at work. … That circulating blood cells can communicate with the host is a chemical reality" (Shapiro 2014; see also Shapiro 2011). For one more testimony, McFall–Ngai et al. (2013) state: "Although animals and bacteria have different forms and lifestyles, they recognize one another and communicate in part because … their genomic 'dictionaries' share a common and deep evolutionary ancestry. … Biologists now know that bacteria have social behaviors, communicating with each other through chemical signaling, such as quorum sensing; more recently, interdomain quorum signaling between bacteria and their eukaryotic partners has become evident" (3232).[10]

Cellular cognition and communication already exist in the wild. Having been around since life got itself going in the form of bacteria some four billion years ago, microbial sentience helps to account for life's evolutionary resilience and inventiveness. While *Blood Music* predates these newer understandings, that this work imagines self-aware, social, inquisitive, and geopolitically ambitious uplifted lymphocytes guessed correctly regarding the actual animation of the terrestrial microcosm. Fifteen years later, both *The Children Star* and *Brain Plague* place intellectual microbes within dynamic mutualistic ecological relations and so transcend apocalyptic ends and other cultural clichés in favor of symbiotic visions of posthuman viability. Properly for its mode, the posthuman comedy of the Elysium Cycle ends with a marriage between the human macrocosm, the machinic mesocosm, and the microbial microcosm.

Notes

1. Matters of scale have been a lively interest in recent critical theory. See Woods (2017), and M. T. Clarke and Wittenberg (2017).
2. See B. Clarke (2017, 141–52). For general bearings on posthumanism in literature, see B. Clarke and Rossini (2017, xi–xxii).
3. The evolutionary thinker Lynn Margulis writing with Dorion Sagan notes, "A legacy of life is to literally incorporate more and more of its environment into itself. … Technology, in short, is an integral part of the ancient ecological cycles of procurement, removal, and reuse that appeared on Earth long before our ancestors turned human" (Margulis and Sagan 2007, 81, 83).
4. The foregoing summary is largely based on Bal (2017). See also B. Clarke (2008, 20–35), and Herman (2009, 105–36).
5. *Daughter of Elysium*'s more sprawling and episodic style makes it the section of the Elysium Cycle that most anticipates the discursive manner of *The Highest Frontier* (2011).
6. The phenomenon of cybernetic self-organization in *Daughter of Elysium*, together with the significant supporting character Lem Inashon, appears

to posit Stanislaw Lem's *Cyberiad*, a posthuman comedy of the first order, as a significant intertext for the Elysium Cycle. The apposite episode from that text is the tale of Mymosh the Selfbegotten: "something emerged at the edge of the dump, not far from the puddle which had by now dried up, and this something, a creature of pure accident, was Mymosh the Selfbegotten. … Mymosh … went flying into the nearby puddle, where his chlorides and iodides mingled with the water, and electrolyte seeped into his head and, bubbling, set up a current there, which traveled around and around, till Mymosh sat up in the mud and thought the following thought:—Apparently, I am!" (Lem 1985, 233–34). For more on the cybernetic posthuman in *The Cyberiad*, see B. Clarke (2008, 107–26).

7. Doggie, with her "beetlelike … polished shell" (1993, 9), would appear to be a distant descendant of cybernetician Grey Walter's robot tortoises from the late 1940s, which had actual metal carapaces. Walter referred to them "as members of a new inorganic species, *Machina speculatrix*. … The tortoises had two back wheels and one front. A battery-operated electrical motor drove the front wheel, causing the tortoise to move forward; another motor caused the front forks to rotate on their axis, so the basic state of the tortoise was a kind of cycloidal wandering" (Pickering 2010, 43).
8. For more discussion of telepathy in relation to narrative discourse, see B. Clarke (2014, 25–37).
9. Prime numbers constitute a main candidate for a universal alphabet in the search for extraterrestrial intelligence (SETI). A series of prime numbers represents the nth degree of non-randomness. Since the odds against a prime series emerging from a random (non-intellectual) process are "astronomical," its reception from an extraterrestrial source would be a "beacon" indicating the presence of (mathematical) intelligence. Other narratives ranging from *Close Encounters of the Third Kind* (Spielberg 1977) to *Contact* (Zemeckis 1997) have also popularized this SETI trope.
10. The newer biological understandings to which Shapiro refers concern how "Carl Woese and George Fox opened a new research frontier by producing sequence-based measures of phylogenic relationships, revealing the deep evolutionary history shared by all living organisms. This game-changing advance catalyzed a rapid development and application of molecular sequencing technologies, which allowed biologists for the first time to recognize the true diversity, ubiquity, and functional capacity of microorganisms" (McFall-Ngai et al. 2013, 3229).

References

Bal, Mieke. 2017. *Narratology: Introduction to the Theory of Narrative*, 4th ed. Toronto: University of Toronto Press.

Bear, Greg. 1985. *Blood Music*. New York: Open Road, 2014.

Bouttier, Sarah. 2018. The 'Right' Amount of Agency: Microscopic Beings vs Other Nonhuman Creatures in Contemporary Poetic Representations. *Épistémocritique* 17 (May). https://epistemocritique.org/the-right-amount-of-agency-microscopic-beings-vs-other-nonhuman-creatures-in-contemporary-poetic-representations/. Accessed 28 Sept 2019.

Clarke, Bruce. 2008. *Posthuman Metamorphosis: Narrative and Systems*. New York: Fordham University Press.

———. 2014. *Neocybernetics and Narrative*. Minneapolis: University of Minnesota Press.

———. 2017. The Nonhuman. In *The Cambridge Companion to Literature and Posthuman*, ed. Bruce Clarke and Manuela Rossini, 141–52. New York: Cambridge University Press.

Clarke, Bruce, and Manuela Rossini (eds.). 2017. *The Cambridge Companion to Literature and the Posthuman*. New York: Cambridge University Press.

Clarke, Michael Tavel, and David Wittenberg (eds.). 2017. *Scale in Literature and Culture*. New York: Palgrave Macmillan.

Herman, David. 2009. The Third Element; or How to Build a Storyworld. In *Basic Elements of Narrative*, 105–36. Hoboken, NJ: Wiley-Blackwell.

Lem, Stanislaw. 1985. *The Cyberiad: Fables for the Cybernetic Age*, trans. Michael Kandel. New York: Harvest.

Margulis, Lynn, and Dorion Sagan. 2007. *Dazzle Gradually: Reflections on the Nature of Nature*. White River Junction, VT: Chelsea Green.

McFall-Ngai, Margaret, et al. 2013. Animals in a Bacterial World, a New Imperative for the Life Sciences. *Proceedings of the National Academy of Sciences* 110 (9) (February 26): 3229–36.

McGurl, Mark. 2012. The Posthuman Comedy. *Critical Inquiry* 38 (3 Spring): 533–53.

Pickering, Andrew. 2010. *The Cybernetic Brain: Sketches of Another Future*. Chicago: University of Chicago Press.

Shapiro, James A. 2011. *Evolution: A View from the 21st Century*. Upper Saddle River, NJ: FT Press Science.

———. 2014. E-mail Message to Author (December 27).

Slonczewski, Joan. 1986. *A Door into Ocean*. New York: Tor Books.

———. 1993. *Daughter of Elysium*. New York: William Morrow.

———. 1998. *The Children Star*. New York: Tor.

———. 2000. *Brain Plague*. New York: Orb.

———. 2011. *The Highest Frontier*. New York: Tor.

Spielberg, Steven, dir. 1977. *Close Encounters of the Third Kind*. Columbia Pictures.

Suvin, Darko. 1979. Estrangement and Cognition. In *Speculations on Speculation: Theories of Science Fiction*. 2005, ed. James Gunn and Matthew Candelaria, 23–35. Lanham, MD: Scarecrow Press.

Vint, Sherryl. 2010. Animal Studies in the Era of Biopower. *Science Fiction Studies* 37 (3 November): 444–55.

Woods, Derek. 2017. Epistemic Things in Charles and Ray Eames's Powers of Ten. In *Scale in Literature and Culture*, ed. M. T. Clarke and Wittenberg, 61–92.

Zemeckis, Robert, dir. 1997. *Contact*. Warner Brothers.

CHAPTER 3

A Door into Ocean as a Model for Feminist Science

Christy Tidwell

> She had seen too many visions of heaven, too many hells, to choose among them.
> I hope we make our own.
>
> —Joan D. Vinge, *The Snow Queen* (1980)

Scientific degrees, jobs, and research have long been dominated by men, and STEM (science, technology, engineering, and math) fields are perceived as masculine. I teach at a STEM school where the student body is nearly 80 percent male, and my students can readily list examples of male scientists (both fictional and nonfictional): Albert Einstein, Dr. Frankenstein, Richard Feynman, Doc Brown, Carl Sagan, Dr. Jekyll, Isaac Newton, Charles Darwin, Nikola Tesla, Stephen Hawking, Alan Turing, and so on. But when I ask them to name female scientists, they can rarely list more than one: Marie Curie. Feminist science fiction like *A Door into Ocean* (1986) provides a welcome corrective to this state of affairs and the perceptions that bolster it. Despite the thirty years that

C. Tidwell (✉)
English and Humanities, South Dakota School of Mines & Technology, Rapid City, SD, USA

© The Author(s) 2020

B. Clarke (ed.), *Posthuman Biopolitics*,
Palgrave Studies in Science and Popular Culture,
https://doi.org/10.1007/978-3-030-36486-1_3

have passed since the publication of *A Door into Ocean*, unfortunately, these disparities remain, and Slonczewski's novel is as politically relevant and necessary now as it was in 1986.

Skidmore (1993) reveals the relationship between gender and science that existed in the mid- to late 1980s: "while women earned 46 percent of bachelor's degrees in mathematics and 45 percent of those in life sciences, they earned less than a third of the degrees awarded in physical science (31 percent), computer science (31 percent), and environmental science (25 percent)" (46). Furthermore, these numbers were not representative of the number of women who went on to earn higher degrees in those fields: in 1990, women were awarded only 24 percent of doctoral degrees in chemistry, 18 percent in mathematics, and 11 percent in physics (46). This was the result of women's lack of funding, which led to their having less research time and fewer opportunities to work with others in their fields, as well as of covert and overt sexual harassment (46). Fewer advanced degrees in these fields meant fewer women working in these fields, too: "Women are under-represented in science and engineering compared to their participation in the U.S. work force (45 percent). In 1988 women comprised 16 percent of all employed scientists and engineers—30 percent of scientists; 4 percent of engineers" (48). Women scientists and engineers were also more consistently unemployed ("at the master's level, 2.7 versus 1.5 percent; and 1.7 and 0.6 percent for Ph.D.s"), underemployed ("Women scientists and engineers were three times as likely as men to report being underemployed in 1986: 6.3 percent versus 1.9 percent"), and underpaid (with yearly earnings at "approximately three-fourths those of men's" [48]). This was the situation when Dr. Joan Slonczewski, then working as an assistant professor of biology, published *A Door into Ocean*.

Sadly, women's position in STEM fields does not look much different today. *The Chronicle of Higher Education* points out that

> Women have made gains in the life sciences, where they receive some 70 percent of bachelor's degrees, yet they hold only 38 percent of assistant professorships in the field, and only 24 percent of the full professorships. Underrepresentation is starker elsewhere: in physical science, technology, engineering, and math, it's not uncommon for women to constitute a tenth of full professorships, and the ratio of American women earning doctorates in these fields peaked at 28 percent in 2009. It has been declining since. (Voosen 2016)

Once women graduate with these degrees, inequities continue—the same kinds of inequities that existed in the 1980s. Furthermore, one study indicates that while women are disproportionately performing the experimental work—the labor—of science, men are more likely to receive credit for the intellectual work such as writing articles and designing experiments (Sugimoto and Lariviere 2016). This indicates that leadership roles are still dominated by men and that women's work and voices are overlooked.

Over the years, feminist science fiction has developed a number of responses to these inequities—including rejecting science, attempting to control it, and embracing it—and, in doing so, it has also explored the possibility of a feminist science. First, some feminist science fiction rejects traditional science in favor of an essentialist, mystical communion with nature that argues all science is masculinist and therefore bad. This is best illustrated by 1970s and 1980s feminist utopias such as Sally Miller Gearhart's *The Wanderground* (1978), Dorothy Bryant's *The Kin of Ata* (1976; first published in 1971 as *The Comforter*), and Judy Grahn's *Mundane's World* (1988), in which the critique of traditional science leads to its abandonment in favor of magic, pseudoscience, or what Roberts (1993) has called "alternative science" (121). These texts privilege personal, revealed knowledge over testable or shared knowledge: *The Kin of Ata* focuses on dreams and embraces the non-rational, while *Mundane's World* emphasizes a mystical ecopsychology that restores connection with the earth (developed in contrast to the activities of science and medicine, which Grahn sees as separating practitioners from nature). These authors also invent imaginary powers to replace the accepted forces of western science. In *The Wanderground*, such inventions take the form of *mindstretch* (a form of psychic communication that extends not only between humans but also between human and nonhuman beings) and the *lonth*, an embodied sense that allows the women to fly. However, despite Roberts' argument that feminist SF recovers magic and witchcraft for feminist purposes and that this recovery "mak[es] women doubly powerful as they claim science for themselves and resist the narrow definition of science offered by patriarchal society" (1993, 8), this move away from traditional science is troubling. Necessary though these critiques of masculinist science may be, their rejection of science or redefinition of science to include magic take it too far. As Squier (2004) notes, "Alliances with nonfeminist scholars in social science and science fields are more readily cemented if the feminist scholar positions herself or himself as sharing the same commitment to *facts* as opposed to *fictions*" (43). Therefore, the rejection of science and adoption of magic in feminist SF potentially closes off the possibility of productive connections with other fields and with nonfeminists.

Another approach to science within feminist science fiction—including Joanna Russ's *The Female Man* (1975), Pamela Sargent's *The Shore of Women* (1986), and Sheri S. Tepper's *The Gate to Women's Country* (1988)—rejects not science itself but the version of science in which men are in control. These texts reverse the traditionally gendered power structure and represent the possibility of women directing science themselves. They ask not just whether women can do science but even whether women might not do it differently—or better. Would a women's science be less sexist or more egalitarian? Would bringing more women into science affect the kinds of science that are done? Ultimately, because these texts' women-run societies are founded on violence and oppression, as are the male-run societies they attempt to replace, Russ, Sargent, and Tepper show that putting science in women's hands doesn't necessarily make it better. In *The Female Man*, Russ provides the utopian vision of Whileaway, in which women do science and run their own lives without the interference of men, but contrasts this with the future of Jael, where men and women are at war; here, Jael and the Womanlanders are scientists, but their scientific efforts are focused on developing plagues and weapons to fight the Manlanders. Sargent's *The Shore of Women* and Tepper's *The Gate to Women's Country* both build worlds in the aftermath of a science abused by men—"[m]en nearly destroyed our world" (Sargent 1986, 178) and "[t]hree hundred years ago almost everyone in the world had died in a great devastation brought about by men" (Tepper 1988, 301)—and shift power and scientific knowledge into the realm of women. As in *The Female Man*, however, this is not simply positive. Women's greater social control is largely gained and kept *through* science (using technology to control men and limit their education, reproductive abilities, etc.), and, as a result, science is central but not liberatory. Science is a tool; in men's hands in the past, it has been harmful, and in women's hands in these novels, it is also harmful. Men damaged and destroyed the earth and oppressed women; women protect the earth but oppress men. So, although these narratives may seem at first to argue for the necessity of putting women in charge, their reversals present a significant criticism of the current system, indicating that domination by one half of the population will never be the answer, no matter which half it is. They illustrate the ease with which women in science can find themselves using the same language and oppressive mechanisms as men have done (as in Tepper and Sargent) and also the limitations of simply removing men from the equation (as in Russ's Whileaway).

A third significant approach to science within feminist science embraces rather than rejects science; it also challenges the conception of science and technology as definitively male without simply reversing the terms of an unequal power structure. This feminist science fiction (which both co-exists with these other texts and extends into contemporary science fiction) represents women taking part in science, and this kind of representation is vital to the project of building a better, more inclusive future science. Examples of this approach include Kate Wilhelm's *The Clewiston Test* (1976), Anne McCaffrey's *Dinosaur Planet* (1978) and *Dinosaur Planet Survivors* (1984), Janet Kagan's *Mirabile* (1991), and Mary Robinette Kowal's Lady Astronaut series in *The Calculating Stars* (2018a) and *The Fated Sky* (2018b), as well as short fiction such as Cathy Hinga Haustein's "Earth and Sky Words" (1990), Nancy Kress's "Computer Virus" (2002), and Marissa Lingen's "The Grandmother-Granddaughter Conspiracy" (2009). By telling stories of women doing science (e.g., xenobiology in McCaffrey and Lingen, genetics in Kagan and Kress) and acknowledging the challenges they face (e.g., overt sexism in *The Clewiston Test*, internalized societal pressure to live up to gendered expectations in "Earth and Sky Words"), feminist SF can challenge gendered stereotypes of scientists and provide new narratives for girls and women to use as either models or warnings. On the one hand, representing female scientists in science fiction does not necessarily mean these representations are feminist, nor does it necessarily describe a feminist science in any way, and it is important to be wary of conflating the inclusion of women with feminism. On the other hand, given the dearth of women in science, it is quite possible that, in our current culture, simply imagining a woman scientist can be a feminist act. However, this mode of feminist SF often simply inserts women into traditional science without questioning its history or its terms. These texts can therefore be valuable in challenging stereotypes about women doing science, but they have little to say about feminist science specifically. Furthermore, these novels do not significantly change the way that women talk about science itself. Thus, including women in science will not necessarily change the fundamental narratives and metaphors of science, just the bodies of those who promote and reproduce them.

A Door into Ocean (1986) is one of the works of feminist science fiction that neither rejects science as a way of knowing and understanding the world nor ignores feminist critiques of science.[1] This novel illustrates the possibility of a feminist science that is not built on femaleness or femininity, one that does not simply invert the power structure or leave the structures of science unchanged. Instead, it provides a vision of a productive and, more importantly, realistic feminist science, defined by three major characteristics: (1) a recognition of women's scientific contributions both now and in the past, within the realm of traditional science and also within often overlooked indigenous and citizen sciences; (2) challenges to dichotomies and hierarchies—for instance, between internal and external, self and other, nature and culture, rational and emotional, male and female—that traditional science may not readily or easily question; and (3) explicit and thorough consideration of the political and ethical ramifications of its choices, narratives, and definitions.[2]

This is not to say that traditional science is completely in opposition to these characteristics. Despite popular conceptions of science as strictly objective and unconcerned with the ethics of its discoveries, many working scientists do acknowledge that science is subjective, creative, and has ethical consequences. Stephen Jay Gould, for instance, writes,

> I criticize the myth that science itself is an objective enterprise, done properly only when scientists can shuck the constraints of their culture and view the world as it really is. ... Science, since people must do it, is a socially embedded activity. It progresses by hunch, vision, and intuition. (qtd. in Harding 1991, 145)

Gould argues that actual science is not, as so many believe, purely logical and objective. Moreover, Zachary Pirtle, a program analyst at NASA, highlights the centrality of creative thinking in scientific progress: "Real science still doesn't work in the strictly deductive way that [Sherlock] Holmes describes, for the best scientific questions, there are no straightforward answers, and a lot of the hard work comes from simply trying to imagine new possibilities" (qtd. in Britt 2010, 67).

Meanwhile, many people believe that scientists are concerned not with the results of their experiments on the world and on others but only with abstract knowledge, and there have certainly already been far too many abuses of power and knowledge in the name of science, including Tuskegee, atomic testing, and the treatment of Henrietta Lacks, to easily set these concerns aside. However, to think that all scientists are equally unconcerned with the consequences of their work would be inaccurate. There are scientists who actively stand against the use of their work in unethical or harmful ways, and they should be acknowledged and supported. Neuroscientist Curtis Bell is one of these scientists. He describes a movement encouraging neuroscientists to approach their field more ethically by opposing military use of their research. Bell and those who signed his pledge take responsibility for the potential consequences of their work, and, although they recognize that this pledge in and of itself will not prevent misuse of neuroscientific research, they do what they can to "help make such applications less acceptable" (Bell 2010).

Therefore, because science as it stands can be both creative and ethical, a specifically feminist science requires not a wholesale rejection of science as we know it but instead a shift in focus to uncover and highlight the feminist or feminist-friendly elements that already exist. Feminist science fiction in general and *A Door into Ocean* in particular help provide this shift in focus. Feminist science fiction can emphasize and endorse elements of traditional science that already exist and that are compatible with feminist goals while simultaneously using the genre's speculative nature to explore the possibility of a feminist science that does not yet fully exist. It can reflect the reality of contemporary scientific practice and also imagine new practices into existence.

A Door into Ocean sets the stage for this complex reflection by placing its narrative on two neighboring planets: Valedon has a culture much like our own (hierarchical, militaristic, capitalistic), while Shora is an all-female ocean world whose citizens have adapted both biologically and socially to its constraints and opportunities. Valan culture is built on stone both symbolically and literally (city names, for example, highlight this: Chrysoport, Pyrrhopolis, Dolomoth, Iridis), reflecting the value they place on strength; it also has a history of removing groups that get in the way of progress—for instance, trolls, "Valedon's extinct native race of anthropoids" (17)—analogous to the genocide of Native Americans in the USA and reflective of the Valans' desire to conquer their world. In contrast, Shorans, or Sharers, have adapted to the demands of a watery world, both culturally and physically, with inner eyelids for seeing underwater and webbed fingers to make swimming easier. They live on rafts on the vast and constantly changing ocean and have no safe places to retreat to when the ocean becomes dangerous, so, in order to survive, they must adapt to their world's ecosystem, learn its patterns, and work to provide for themselves while avoiding major disruptions of those patterns. Sharers do not attempt to conquer the world but to live with it. The primary tension of this novel regards the interaction between these two worlds. While the Sharers would gladly live peaceably as neighbors, Valans see Shora as nothing more than a resource: "Seasilk and minerals—that was what Shora meant" (27). As these tensions rise, individuals from Shora and Valedon visit each others' worlds, the Sharers attempting to share understanding with the Valans and the Valans attempting to gain access to the resources they want.

The Sharers' way of life points to the element of feminist science that involves re-examining the history of scientific knowledge and, instead of ignoring the work done by women outside of laboratory settings, reconsidering what counts as science to include this work. Such a redefinition would allow for the inclusion of amateur science, the kind of science that women have historically been more able to do (e.g., midwifery, herbal, and plant knowledges). This inclusion could lead to dramatic changes in the way science is defined, both historically and currently, "from what men do to what *people* do" (Stanley 1983, 5). As an illustration, Lederman (2001) posits an alternate vision of the beginnings of science:

> Imagine an individual in a pre-Scientific Revolution society posing herself the following questions: "Does the extract of the purple-leaved plant which grows on the hill a morning's walk toward the sun reduce the redness when applied to the wound caused by the bite of the spider? Does the liquid made by boiling the leaf stop the shivering of the fever which comes every year when the sun is low in the sky? If it stops the shivering, could the boiled leaves of the plant which grows by the river stop the shivering as well? The leaves of the plant by the river are green, but they are the same shape as the leaves of the purple plant." For all we know, a train of thoughts similar to this may have been one of the starts of science. It contains all the components of science as we know it today; actually, it combines aspects of both experimental medicine … and biological classifications. (441)

Excluding this kind of work from our definitions of science both builds upon and reinforces the current vision of science as separate from the lived world, something that one only does in a sterile environment and/or academic setting; however, the methods involved in Lederman's example and in traditional science are essentially the same. Redefining science to include such activities allows for a broader history of scientific knowledge that includes more of the achievements of women (and indigenous peoples) and also provides a vision of science that does not marginalize or exclude women. As Sandra Harding writes, "It is unclear how one would define this term (scientific method) in such a way that highly trained scientists and junior members of research teams in physics counted as scientists, but farmers in simple societies (or mothers!) did not" (Harding 1987, 28). In this broader definition, women are and have long been scientists, too, even when they have not been welcomed into academic or laboratory settings.

Placing the Sharers' scientific work on the rafts where they live engages in and illustrates this project of redefinition by bringing together lived experience and laboratory science. Because they both live and work on the rafts, protecting their homes also amounts to protection of their scientific knowledge: "Sharers maintained libraries of genes for many species, from edible fish and weeds to seaswallowers and shockwraiths. … The ultimate library was kept within raftwood: every living cell of every raft held a library within its genes, millions of units within a cell too small to see" (249). The two parts of their lives are not easily divisible. Donawerth (1997) argues that the novel develops "a utopian science that puts science in the home, that reconceives the home as organically related to the natural environment, and that thus is controlled by an ethic very different from our current science—a sharing ethic" (12).

Of course, bringing science into the home is not inherently positive, as shown by historical evidence such as the birth control studies done in Puerto Rico:

> Having the women take the Pill at home not only reduced the institutional cost of the trials but also placed the subjects within the domestic context of ordinary life, thus extending the scope of the trial outside medical institutions: every private home could potentially become an experimental site. The El Fanguito housing complex became an externalized and extended domestic pharmaceutical laboratory. (Preciado 2013, 188–89)

This project extended corporate and scientific control into private lives by using "housing modernization as a way of installing a micropharmaceutical laboratory within the domestic environment" (Preciado 2013, 189).

However, the Sharers are not inviting outsiders into their homes and giving them control; they themselves control their homes/laboratories, the lives lived there, and the science done there.

The model of feminist science provided by *A Door into Ocean* thus presents a scientific practice that is built upon scientific principles and values of experimentation and rationality but that also, in its emphasis on the organic (e.g., gene libraries, close ties between home and laboratory), recognizes the importance of the natural world and places the scientist within that world rather than above or outside it. This is a direct challenge to binary, hierarchical models of scientific practice. As feminist theorists of science have argued, traditional (masculine) science has too often either created or reinforced a separation between the scientist and the world being studied. Further, a survey of women's attitudes toward and beliefs about science conducted by Jean Barr and Lynda Birke finds that

> [m]etaphors of control and manipulation abound, almost half of the returns (46 percent) referring explicitly to science's and scientists' isolated, "separate," and abstract(ed) qualities—its/their "dissociation from reality," in the words of one woman. One respondent clearly sees laboratories as *removed* from nature when she writes: "works in a lab far away from nature." (1998, 123)

The Sharers' embodied and embedded approach to science offers a direct counter to such perceptions of science, providing a productive relationship between the scientists and their subjects as well as a more welcoming vision of science.

Moreover, the Sharers' libraries of genes are not simply for abstract knowledge but are also part of their responsibility "to share care for all the lesser sharers as for themselves" (249); thus, caring for their work means caring for themselves and for the larger world. The Sharers therefore enact the argument in Preciado (2013) that minority knowledges must be saved by transforming them

> into collective experimentation, into physical practice, into ways of life and forms of cohabitation. … We must reclaim the right to participate in the *construction* of biopolitical fictions. We have the right to demand collective and "common" ownership of the biocodes of gender, sex, and race. We must wrest them from private hands. (349, 352)

The Sharers are certainly anti-corporate and anti-capitalist; they resist the Valans' efforts to mine their planet for profit and hold the planet and its resources in trust—not just for themselves but also for those without a voice. Their gene libraries reflect this kind of "collective and 'common' ownership of the biocodes" of Shora as well. Most importantly, however, they are actively participating in "the *construction* of biopolitical fictions" by being willing to modify themselves to fit the planet rather than simply trying to shape the planet to match their own desires. This leads to a practice that "will consist of a position, responsible corporal political practice, so that anyone wishing to be a political subject will begin by being the lab rat in her or his own laboratory" (Preciado 2013, 353). Sharers are embodied political subjects—their own lab rats, but no one else's.

This placement of the laboratory in the home is just the first of many challenges the novel presents to traditional dichotomies, including the binary conception of gender. At first glance, *A Door into Ocean*, with its utopian all-female world, seems to rely upon the same kind of separatism present in many other feminist science fiction novels (such as *The Wanderground*, *The Female Man*, *The Gate to Women's Country*, and *A Shore of Women*), but instead uses the conceit of the separatist all-female world to break down gender binaries. The two worlds are built upon very different values that might be read as representing fundamental gender differences (militaristic versus peaceful, hierarchical versus communal); however, many of the central characters belie this simplistic division. For instance, some female characters are more at home with Valan values than with Shoran ones. Jade, for example, a member of the Valan military, proves herself to be hard, cruel, and heartless; her femaleness is far less important to her character than her Valan-ness. On the other hand, Realgar, the military leader who fights the Sharers throughout the book, is eventually changed by the Sharers' ideas and comes to see the world differently: "Somehow, he would never see a glass again, or look into the eyes of a cornered bear, without knowing that the wildest thing he ever hunted still swam beyond his grasp" (365). Ultimately, as a result of these changes, he is unable to betray the Sharers. He even tells his leader, "If every planet in the Patriarchy refused to be ruled, *we all would be free*" (363), which is a very Shoran response and one that leads his leader to express serious disappointment in him as someone who "listened to *them* for too long" (363). Even more dramatically, Spinel, a Valan boy, learns the Sharers' techniques and chooses to live on Shora even with no hope of returning to his home on Valedon. Many Sharers were initially skeptical of Spinel, as a Valan and a "malefreak," but eventually they come to see that Spinel can be both male and Sharer, that Shoran does not only equal woman. Nor is Spinel's transformation incidental to the book's trajectory; instead, it is the ultimate illustration of the power of the Shoran utopia, which is that their ideas about nonviolence, environmentalism, and interconnectedness are transmissible.

The Sharers' language itself helps spread these ideas, further challenging binary modes of thinking as well as emphasizing the political and ethical consequences of linguistic and narrative choices. The structure of the Valan language, although more oriented toward stone, is much the same as ours, and the language includes few unfamiliar words or phrases. In contrast, Sharer language is fundamentally different from ours. As their name indicates, the Sharers' language—and mode of thought—is built upon sharing. Merwen, visiting Valedon early in the book, describes the concept of "wordsharing" to Spinel by reminding him that "[e]ach force has an equal and opposite force" (34). Their language highlights this action and reaction by using "share-forms" like "learnsharing, worksharing, lovesharing" (34). Spinel finds this concept difficult and tests it against Merwen's patient responses:

> "Do you say 'hitsharing,' too? If I hit a rock with a chisel, does the rock hit me?
>
> "I would think so. Don't you feel it in your arm?"
>
> He frowned and sought a better example; it was so obvious, it was impossible to explain. "I've got it: if Beryl bears a child, does the child bear Beryl? That's ridiculous."
>
> "A mother is born when her child comes."
>
> "Or if I swim in the sea, does the sea swim in me?"
>
> "Does it not?"
>
> Helplessly, he thought, She can't be that crazy. "Please, you do know the difference, don't you?"
>
> "Of course. What does it matter?" (34)

This emphasis on sharing begins but does not end with language; it is the way the Sharers see the world. If all is shared, if all action is both given and received at one and the same time, then the Sharers' entire identity is founded on responsibility to their world. Their language highlights how preservation of Shora is also preservation of themselves. They share world preservation. Therefore, environmental care is not simply a sort of stewardship but a relationship of equals, among equals. After all, Merwen asks, "who rules without being ruled?" (34). As Vint (2010) writes,

> Sharers use their sharing verbs for their relationships with all species. This incorporates other species into their language and its ways of structuring the world, suggesting a model of subjectivity that is quite other than that of Western metaphysics. … The Sharer language and social order is premised on the capacity of other species to respond, not merely react. Thus, the Sharers are … involved in a dynamic relationship with other species on their planet that they speak with—not speak for or speak instead of. (Vint 2010, 448)

By highlighting the interconnectedness of self and other, the Sharers' use of language demonstrates how important language and metaphor are to constructing fair and ethical practices, including within the sciences.

This attention to language and metaphor in the sciences echoes and is echoed by the work done within feminist science studies. Spanier (1995) argues that there is bias in the very language used to do science that affects both who engages in science and what science is done. Spanier shows how deeply gender ideologies are embedded in science itself, arguing that the gendering of nongendered elements of science often creates a hostile environment for women. For instance,

> When scientists look to nature, they often bring with them their sociopolitical beliefs about what is natural. This self-reinforcing and internally consistent process creates, reflects, and reinscribes unquestioned assumptions about our world. Stereotypic attributes and behaviors, such as aggressive hunting and fighting versus passive coyness, are superimposed onto animals, often through culturally distorted language. Calling several females with a single male a "harem" conjures up quite a different power relationship from what is now called the "matriarchal" organization of elephants. (20)

Martin (1991) illustrates this further in a now classic study of the language used to describe the process of conception. She describes the way that textbooks presented the egg as passive and the sperm as active, despite evidence contradicting this narrative, and notes that this "imagery keeps alive some of the hoariest old stereotypes about weak damsels in distress and their strong male rescuers. That these stereotypes are now being written at the level of the *cell* constitutes a powerful move to make them so natural as to be beyond alteration" (500). Ultimately, she argues,

> One clear feminist challenge is to wake up sleeping metaphors in science. … Although the literary convention is to call such metaphors "dead," they are not so much dead as sleeping, hidden within the scientific content of texts—and all the more powerful for it. Waking up such metaphors, by becoming aware of when we are projecting cultural imagery onto what we study, will improve our ability to investigate and understand nature. Waking up such metaphors, by becoming aware of their implications, will rob them of their power to naturalize our social conventions about gender. (501)

Part of a feminist intervention in the sciences, then, must involve critique of the narratives and metaphors we already rely upon. Waking up such metaphors will make science a friendlier place for women and feminism. Reminiscent of Haraway's (1984) call for "pleasure in the confusion of boundaries and for responsibility in their construction" (8), we must be not only mindful of the metaphors we use but responsible for the metaphors we create.

The Sharers' use of metaphor illustrates this kind of responsibility. As opposed to what is shown in Martin's study, the objects of study are not unnecessarily gendered, but they are also not even really simply *objects* of study, for that conception would run counter to the mutuality inherent in the Sharers' language and mindset. For instance, they do not see other species, even species they rely upon for communication (e.g., clickflies) as objects; instead, they "share care for all the lesser sharers as for themselves" (267).

This use of language leads to the third element of *A Door into Ocean*'s model of feminist science: the Sharers' emphasis on the ethics of scientific practice. Developing an ethically grounded science and worldview is not unique to feminist science, but it is necessary for any science that claims to be feminist. In other words, science that is not explicitly feminist may also concern itself with the consequences of its actions, but no science without such concern can be plausibly feminist.

Slonczewski's vision of an ethical, feminist science is particularly significant because—as opposed to the approach taken in texts like *The Female*

Man, *The Shore of Women*, and *The Gate to Women's Country*—it consistently relies upon nonviolence. It does so not as a way of reinforcing gender divisions, by implying that the Sharers are nonviolent simply because they are women, but as an active and continually renewing process of self-discipline. The Sharers' use of whitetrance, a deep meditative state, to conduct sit-ins at the Valans' rafts illustrates this discipline. They gather at the edge, enter whitetrance, and refuse to leave. On some occasions, the Valan soldiers push them into the water while they are in this state, but they never respond with violence or anger toward the Valans. Some of the women must work to maintain a nonviolent posture, wishing to fight back and feeling tempted to use violence. The Shoran leaders, however, know that this is unwise and dissuade the angry and frightened Sharers from deserting their nonviolent principles. In fact, in the one instance where violence is used by a Shoran, when Nisi attacks the Valan base, her action prompts some of the most direct violence from the Valans. Realgar, the Valan leader, believes that Nisi must have been supported by other Sharers because "Sharers never act alone. All decide for one, and one for all" (307). He is wrong about this, but it does not stop him from using this as an excuse to institute a new policy: "From now on, the slightest infraction of Valan orders will be met with execution on the spot. To demonstrate our intent, the Protector and four of her Councillors shall be put to death" (308).

Despite this complication, the Sharers' nonviolence works. The Sharers' nonviolent ways are able to fend off the Valan invaders. It may be temporary, but it is a concrete victory. This consistent rejection of violence in the creation and defense of Shora and the success of nonviolence methods provides a clear endorsement of the Sharers' approach. In the face of a history in which science has often been used to further violence or control others, this vision of science without domination is promising and a necessary part of a truly feminist science that must include consideration of the political and ethical ramifications of its choices, narratives, and definitions. The Sharers' ethic of nonviolence, applied to scientific practice, is one way of considering such ramifications. *A Door into Ocean* is a valuable model for feminist science in repeatedly highlighting the role that we as individuals and as a culture have to play in creating a world where this kind of science exists. Judith Merril writes,

> Some people, and I am one, also believe that art is by nature revolutionary: that a vital function of the artist is to produce and publish "virtual realities"

> of social change. Certainly the inverse is true: no radical change can ever occur until a believable and seductive new vision is made public. Professors and politicians may seduce, but only artists can create belief in the new vision—the new myth. (Merril and Pohl-Weary 2002, 42)

A Door into Ocean provides such a seductive vision of a feminist science. Of course, its author does not bear the weight of creating this better future alone—none of us do. Instead, all of us who are concerned with science and with gender equality share this project of creating a better science and a better world.

Notes

1. Other feminist SF texts that do similar work include Le Guin (1974), Piercy (1976), Griffith (1992), Hegland (1996), and Gloss (2003).
2. Subramaniam (2014) provides a detailed description of the way these dichotomies and hierarchies shape science as a discipline: "The 'culture of no culture' turns out to be entirely about culture, identity, and difference, as does scientific knowledge" (23).

References

Barr, Jean, and Lynda Birke. 1998. *Common Science? Women, Science, and Knowledge*. Bloomington: Indiana University Press.

Bell, Curtis. 2010. Neurons for Peace. *New Scientist* 205 (2746): 24–35.

Britt, Ryan. 2010. Sherlock Holmes and the Science Fiction of Deduction. *Clarkesworld Magazine*, November.

Bryant, Dorothy. 1976. *The Kin of Ata Are Waiting for You*. New York: Random House.

Donawerth, Jane. 1997. *Frankenstein's Daughters: Women Writing Science Fiction*. Syracuse: Syracuse University Press.

Gearhart, Sally M. 1978. *The Wanderground: Stories of the Hill Women*. London: Persephone Press.

Gloss, Molly. 2003. Lambing Season. In *The Year's Best Science Fiction: Twentieth Annual Collection*, ed. Gardner Dozois, 135–44. New York: St. Martin's Press.

Grahn, Judy. 1988. *Mundane's World*. Freedom, CA: Crossing Press.

Griffith, Nicola. 1992. *Ammonite*. New York: Del Rey Books.

Haraway, Donna. 1984. A Manifesto for Cyborgs: Science, Technology, and Socialist Feminism in the 1980s. In *The Haraway Reader*, 7–46. New York: Routledge, 2004.

Harding, Sandra. 1987. The Method Question. *Hypatia* 2 (1): 19–33.
———. 1991. *Whose Science? Whose Knowledge? Thinking from Women's Lives.* Ithaca: Cornell University Press.
Haustein, Cathy Hinga. 1990. Earth and Sky Words. In *The Women Who Walk Through Fire: Women's Fantasy and Science Fiction*, vol. 2, ed. Susanna J. Sturgis, 12–22. Freedom, CA: Crossing Press.
Hegland, Jean. 1996. *Into the Forest.* New York: Bantam Books.
Kagan, Janet. 1991. *Mirabile.* New York: Tor Books.
Kowal, Mary Robinette. 2018a. *The Calculating Stars.* New York: Tor Books.
———. 2018b. *The Fated Sky.* New York: Tor Books.
Kress, Nancy. 2002. Computer Virus. In *The Year's Best Science Fiction: Nineteenth Annual Collection*, ed. Gardner Dozois, 106–40. New York: St. Martin's Griffin.
Lederman, Muriel. 2001. Structuring Feminist Science. In *The Gender and Science Reader*, ed. Muriel Lederman and Ingrid Bartsch, 437–46. London: Routledge.
Le Guin, Ursula K. 1974. *The Dispossessed: An Ambiguous Utopia.* New York: Harper.
Lingen, Marissa. 2009. The Grandmother-Granddaughter Conspiracy. *Clarkesworld Magazine*, December.
Martin, Emily. 1991. The Egg and the Sperm: How Science Has Constructed a Romance Based on Stereotypical Male-Female Roles. *Signs* 16 (3): 485–501.
McCaffrey, Anne. 1978. *Dinosaur Planet.* New York: Ballantine Books.
———. 1984. *Dinosaur Planet Survivors.* New York: Ballantine Books.
Merril, Judith, and Emily Pohl-Weary. 2002. *Better to Have Loved: The Life of Judith Merril.* Toronto: Between the Lines.
Piercy, Marge. 1976. *Woman on the Edge of Time.* New York: Fawcett Crest.
Preciado, Beatriz [Paul B.]. 2013. *Testo Junkie: Sex, Drugs, and Biopolitics in the Pharmacopornographic Era.* New York: The Feminist Press.
Roberts, Robin. 1993. *A New Species: Gender and Science in Science Fiction.* Champaign: University of Illinois Press.
Russ, Joanna. 1975. *The Female Man.* Boston: Beacon Press, 1986.
Sargent, Pamela. 1986. *The Shore of Women.* New York: Crown.
Skidmore, Linda. 1993. Women in Science: Past and Future Trends. In *Women at Work: A Meeting on the Status of Women in Astronomy*, ed. C. Megan Urry, Laura Danly, Lisa E. Sherbert, and Shireen Gonzaga, 45–55. Baltimore: Space Telescope Science Institute.
Slonczewski, Joan. 1986. *A Door into Ocean.* New York: Arbor House.
Spanier, Bonnie B. 1995. *Im/Partial Science: Gender Ideology in Molecular Biology.* Bloomington: Indiana University Press.
Squier, Susan Merrill. 2004. *Liminal Lives: Imagining the Human at the Frontiers of Biomedicine.* Durham: Duke University Press.

Stanley, Autumn. 1983. Women Hold Up Two-Thirds of the Sky: Notes for a Revised History of Technology. In *Machina Ex Dea: Feminist Perspectives on Technology*, ed. Joan Rothschild, 5–22. Oxford: Pergamon Press.

Subramaniam, Banu. 2014. *Ghost Stories for Darwin: The Science of Variation and the Politics of Diversity.* Champaign: University of Illinois Press.

Sugimoto, Cassidy R., and Vincent Lariviere. 2016. Perspectives: Giving Credit Where It Is Due. *Chemical & Engineering News* 94 (35 September): 32–33.

Tepper, Sheri S. 1988. *The Gate to Women's Country.* New York: Bantam Books.

Vinge, Joan D. 1980. *The Snow Queen.* New York: Tor Books.

Vint, Sherryl. 2010. Animal Studies in the Era of Biopower. *Science Fiction Studies* 37 (3 November): 444–55.

Voosen, Paul. 2016. The Subtle Ways Gender Gaps Persist in Science. *The Chronicle of Higher Education*, March 6. http://www.chronicle.com/article/The-Subtle-Ways-Gender-Gaps/235598. Accessed 24 Sept 2019.

Wilhelm, Kate. 1976. *The Clewiston Test.* New York: Farrar Straus Giroux.

CHAPTER 4

"Then Came Pantropy": Grotesque Bodies, Multispecies Flourishing, and Human–Animal Relationships in *A Door into Ocean*

Chris Pak

In science-fictional worlds, the colonization and habitation of planets calls for the physical modification of space (terraforming) or of bodies (pantropy). While terraforming is often the preferred method for adapting to the conditions of new worlds, pantropy supplements this planetary modification. Discussions of terraforming link to issues of climate change, while those of pantropy raise issues related to genetic modification. The oceanic world depicted in *A Door into Ocean* (1986) constrains the way adaptation to and modification of the environment can be conceived. Adapted animal and human bodies evoke the monstrous, grotesque, and sublime; grotesque bodies interrogate the meaning of the animal, the human, and nature. Considering these intersections allows us to explore the distinct interventions into nature that terraforming and pantropy entail, and how pantropy critiques colonialist approaches to terraforming. After a brief discussion of terraforming and pantropy, I will explore how Slonczewski uses pantropy to question the values and assumptions that

C. Pak (✉)
College of Arts and Humanities, Swansea University, Swansea, UK

© The Author(s) 2020

B. Clarke (ed.), *Posthuman Biopolitics*,
Palgrave Studies in Science and Popular Culture,
https://doi.org/10.1007/978-3-030-36486-1_4

underlie the pursuit of terraforming. Through connections between individual bodies and the ecological body of the world, pantropy in *A Door into Ocean* refigures terraforming into symbiogenetic communities that adapt and maintain their global environment. The grotesque imagery so produced is fundamental to the text's challenge to colonialist domination embodied in industrial approaches to terraforming. The pantropic subjects and the ecology of the planet Shora offer an alternative conception of habitation centered on responsiveness to other lives.

Jack Williamson coined "terraforming" in his 1942 short story "Collision Orbit" (Blish 2001 [1957], 44). In the same year, James Blish coined "pantrope" to refer to the microscopic humans adapted to inhabit a puddle of water in "Sunken Universe" (Blish 1942), later expanded into "Surface Tension" (Blish 1952) and incorporated into the fixup novel *The Seedling Stars* (Blish 2001 [1957]). The emergence of the terms "terraforming" and "pantropy" in the same year suggests an increasing sophistication in the construction of dialogues about space colonization in the 1940s. While not exclusive of one another, terraforming and pantropy involve two different modes of habitation, with distinct underlying philosophies with regard to the otherness of nature and other civilizations. "Pantropy," loosely translated, means "changing everything" (Blish 2001 [1957], 8). In *The Seedling Stars*, the narrator points to pantropy's mythic resonances, arguing that "it went back, in essence, as far as Proteus—and as deep into the human mind as the werewolf, the vampire, the fairy changeling, the transmigrated soul" (44). This chapter's title, "Then Came Pantropy," is drawn from "Seeding Program" in *The Seedling Stars*, first published as "A Time To Survive" (Blish 1956). Colonialism underlies humanity's approach to space colonization in many terraforming narratives. The opposition between terraforming and pantropy turns on a philosophical choice between adapting the other or adapting the self. Terraforming and pantropy are co-dependent technologies that help to accelerate the habitation of other planets. Since *The Seedling Stars*, the sense of pantropy has expanded to include body modifications other than genetic adaptations, including cyborgization.

In *A Door into Ocean*, the colonizing Valans attempt to establish sovereignty over the indigenous Sharers, in part justifying this endeavor by appealing to the benefits of free trade, and later by pointing to the profits that mineral extraction, Valan fisheries, and seasilk production generate. The adaptation of the environment for the purpose of resource extraction threatens the ecological networks on the planet. The danger that terraforming poses to indigenous communities lies in its disrespect and destruction of modes of habitation based upon a co-adaptation of "amborg" communities and their environments. Gordon (2008) coined "amborg" to "represent the human/animal interface" understood as "organisms in their most liminal states, not just humans when we acknowledge our family tree, but any animals that interact with, exchange glances with, and acknowledge the presence and sentience of another species" (191). The critique of terraforming via pantropy illuminates the choice between the destruction or persistence of existing ecological and cultural networks.

Multispecies Flourishing

A Door into Ocean narrates the struggle between the inhabitants of the ocean planet Shora and occupiers from Valedon, part of the interplanetary Torran empire. Drawing from her Quaker background, Slonczewski presents the indigenous, all-female Sharers as pacifists who practice a consensus-based form of government that stands in stark contrast to the colonizers, who privilege violence and an adherence to hierarchy as strength. In an attempt to preserve their communities, the Sharers Merwen and Usha bring the Valan adolescent Spinel—a "malefreak"—to live with them and the Valan noble, Lady Berenice. Berenice is engaged to Realgar, an officer who later becomes head of the Valan colonizing force. By dwelling with the Sharers, Spinel helps them to understand their occupiers: the relationships he forms provide new contexts for the Sharers to think about what it means to be human, and this in turn assists Merwen in swaying other Sharers from violent reprisals against the Valans. As the narrative unfolds, the Sharers' nonviolent protest and stewardship of their multispecies communities leads them successfully to oppose colonization and terraforming by the Valans.

Exploitation of all the planet's inhabitants push the indigenous people to resist occupation. Vint (2010) explores how *A Door into Ocean* offers an alternative to a form of colonial governance based on submission to Valan sovereignty. The concept of sovereignty is important for terraforming narratives because the notion of the liberal human subject at the base of the Hobbesian social contract informs and justifies colonization and terraforming. The Sharers challenge these discourses of human nature and governance by grounding their society on a concept of the human based on a distributed responsibility to ensure multispecies flourishing, itself enabled by a practice of consensus building. In contrast to the colonizing Valans' hierarchical and power-based system of sovereignty, which privileges the human as the only subject capable of responding and not merely reacting to others, the Sharers "have no concept of sovereignty that coincides with the [Hobbesian] social contract model, and the Valan occupation force seeks in vain to find a leader who can speak for all the people" (Vint 2010, 448). According to Vint, the Sharer language is emblematic of their lack of a concept for sovereignty: its reciprocal verb-forms and lack of subject/object distinction directs readers' attention to the power of naming in shoring up the boundary between the human and animal, which the Valans take as axiomatic. *A Door into Ocean* articulates "a new ethics of passivity and diffuse subjectivity that could be one of the new fables of transformed sovereignty" that Derrida calls for in his work (Vint 2010, 449). This attempt to rethink sovereignty is often presented as a consequence of pantropic adaptations in sf. The Sharers' physical adaptations and unisex population mark them as morphologically distinct from the colonizing Valans, thus underscoring how embodiment aligns with political differences. Pantropy challenges the values that underpin colonial approaches to terraforming, which is aligned in this novel to masculinity, hierarchy, individualism, and violence—key elements of the understanding of human nature upon which the Hobbesian articulation of sovereignty is based.

We can read Slonczewski's world as a model for thinking about contemporary biopolitics and human–animal interactions. Shora is an inversion of the desert world of Arrakis in *Dune* (Herbert 1965), while the rafts the Sharers inhabit invert the forests of Le Guin's Athshe in *The Word for World Is Forest* (1976). In contrast to Herbert and Le Guin, Slonczewski explores the possibilities for nonviolent resistance to a colonizing force. The Sharers colonized their world in the distant past; Spinel's ability to eat food on Shora, to process the DNA of alien organisms, "can only be explained if the Sharers have systematically replaced most, if not all, of the pre-existing ecosystem of Shora with Earth-evolved organisms—in other words, in their own way they must have terraformed Shora—just as Valedon was terraformed" (Slonczewski 2001). Terraforming is thus a part of Shora's past as much as a threat to its future, an issue that Slonczewski explores in more detail in the sequels *Daughter of Elysium* and *The Children Star*.

Dupré (2002, 54–55) argued for a promiscuous realism that recognizes the artifice involved in constructing taxonomies to account for nonhuman bodies. This does not make such classifications illegitimate, but it does call for flexibility when choosing an appropriate mode of classification. As Agamben (2004) states in *The Open*, the notions of human and animal were constructed through the historical disavowal of the animal other: "the division of life into vegetal and relational, organic and animal, animal and human, therefore passes first of all as a mobile border within living man, and without this intimate caesura the very decision of what is human and what is not would probably not be possible" (15). The Sharers's struggle with the question of Valan humanity undermines the Hobbesian social contract and its notion of sovereignty. Spinel's time on Shora is a struggle to overcome the ways he has traditionally classified humans and other animals, substituting instead elements of the Sharers' mode of thinking. The body of the self and other is the first site for the formation of classification; the body in context, its lived experience in contact with other bodies, allows Spinel to revise these classifications and to reconsider the basis of his culture's orientation to nature.

Shora's ecology is complex and well-delineated. The indigenous human inhabitants call themselves and other species Sharers. They live in multispecies communities on rafts, which grow from seeds into large organisms that provide living space for many animals. The relationships that Sharers maintain with the creatures who live around and share their rafts are basic to their survival. Close physical contact with these companions make them amborg subjects who must respond appropriately to the needs of their fellow species. Clickflies, for example, provide the Sharers with the capacity to record and share information relevant to their practice of lifeshaping, for which the cells of their rafts function as an information storage system. The Sharers are both foragers and hunters, offering what Le Guin (1996) has described as a "carrier bag narrative" of survival within the ecological limits imposed by the environment, without prioritizing masculinist narratives. While they depend on the lives of other animals for their existence, this relationship is not one of dominance but of symbiosis. Sharers hunt the deadly shockwraith for their tentacles, accepting the danger their mode of living entails while resisting the extermination of threatening animals, cognizant as they are of how the healthy flourishing of all species depends on ecosystems. This multispecies community is a permaculture that has developed over ten thousand years, constituted of amborg subjects co-evolving and responding dynamically to each other. One example involves the Sharers' use of biomedical technologies. The Sharers rescue Valans who are caught in an explosion and encase them in a living cocoon "that bristled with vines and trailers" (356). While not strictly animals, these organisms connect the multispecies animal communities to their world, functioning as a life-support system that helps the Sharers to flourish in their dangerous environments.

Scientific taxonomy is only one of the ways *A Door into Ocean* represents and orders animals. While it provides a powerful theme connecting life on Shora to that on Earth, it also frames other modes of classification. This includes the Valans' colonial-racist classification of Sharers as "catfish" (a reference to their baldness, streamlined ears, and broad, flat lips), and the Sharer's philosophical system that identifies all as Sharers. There is evidence of other taxonomies, such as when Merwen identifies predators as "blood drinkers" (235) and when Yinevra rejects Merwen's classification of the Valans as human, a claim Merwen bases on their genetic compatibility (32): For Yinevra, shared evolutionary ancestry is insufficient to qualify as human. The novel's critique of colonialism is rooted in dismantling ossified modes of classification, such as the Valans' sense that their idea of their humanity puts them above the Sharers, people and animals. In order to adequately unpack the nature of the critique of terraforming that pantropy represents, it is also important to consider how the grotesque operates in this novel. In the context of multispecies flourishing and amborg subjectivity, bodies are the matrix for emergent entities poised in ever-shifting relationships. Classification is subject to constant flux, necessitating a promiscuous realism that is responsive to embodied context. In *A Door into Ocean*, the Sharers' expertise with lifeshaping and the monstrous physicality of the Shoran ecology—as Spinel experiences it—enables thinking about the sf grotesque and what it can tell us about terraforming, pantropy, and the human relationship to nature.

Haraway (2008) adapts Bowker and Star's notion of "torque" to "describe the lives of those who are subject to twisted skeins of conflicting categories and systems of measure or standardization. Where biographies and categories twine in conflicting trajectories, there is torque" (134). *A Door into Ocean* dramatizes this notion of torque through the conflict between two histories, between Shoran multispecies flourishing and Valedon's imperial subjugation of nature, and between the stories of the adopted Valans Spinel and Berenice, whose experiences of life on Shora offer amborg entanglements that lead to new communities. It is significant that when Spinel first meets Merwen, she responds to his question about a possible Shoran invasion of Valedon by noting that Shora has long been occupied and that "truth is a tangled skein, and time ravels it" (15). *A Door into Ocean* explores the problematic of truth through a narrative that creates spaces for the emergence of amborg subjects.

Through the figure of the sower seeding new worlds, Haraway connects multispecies communities to terraforming as the creation of new contexts for shared lived experiences and new myths for explaining what it means to be human. For Haraway, "sowing worlds is about opening up the story of companion species to more of its relentless diversity and urgent trouble" (2013, 138). This conceptualization of terraforming opens up the notion of flourishing to include the relationships and interactions formed between the biotic and abiotic. Bacteria and plants are points of connection between species and their environments: "symbiogenesis is not a synonym for the good, but for becoming with each other in response-ability" (2013, 145), a symbiotic form of becoming that brings with it an ethical imperative to be responsive to other species. One way in which pantropy critiques Valan colonization is by drawing a distinction between different modes of terraforming, with pantropy positioned as a contrast to destructive modes of planetary adaptation that erase, rather than incorporate, prior ecological systems. Pantropy functions as an emblem for the ongoing story of symbiogenesis. The portrayal of diverse animal others is crucial to this vision because the attenuation of the network of multispecies relationships is presented as a profound impoverishment of both ecological and cultural vitality. Conflict arises from the Valans' failure to respect the pre-existing relationships on Shora. The novel articulates this critique of terraforming as the colonial imposition of systems of classification onto nature through four animals that appear in *A Door into Ocean*: clickflies, breathmicrobes, humans, and seaswallowers.

Clickflies: Companion Species and the Carnivalesque Uncrowning

There are two uses of the word "grotesque" in *A Door into Ocean*: one when Spinel begins to succumb to what the Valans call the "purple plague" and one when he first sees a clickfly. These insects, compared at one point to "bees at a honeycomb" (50), are fundamental to Sharer society, providing a communications network that links raft-communities who would otherwise be isolated for days or weeks, depending on the length of a physical journey. Clickflies can transmit messages and navigate toward other raft-communities, and are thus valued companion species. The Sharers can decode their webs and use this system of "writing" to teach lifeshaping, the most essential of their arts. These webs encode at the genetic level, offering a vast capacity for storage and an ideal basis for the Sharers' advanced biological science—what the Valans fear as a "Forbidden Science" imbued with all the grotesquerie and chimerical monstrosity that (loss of) control over the body has signified since at least Wells's *The Island of Dr Moreau.*[1] The Sharers celebrate clickflies for making their practice of learnsharing possible.

The Valan reaction to the clickflies is clearly a reaction, not a response. Spinel's sense of grotesquerie at the sight of the clickfly is apt, for they demonstrate an element of the carnivalesque uncrowning that leads Spinel—and readers—to a new insight into the Sharers' way of life. Carnivalesque uncrownings are reversals that turn the world on its head (Bakhtin 1984, 79). The transformation of Spinel's orientation toward the clickflies from a feeling of unease to one in which he accepts becoming Shoran is anticipated in the very first episode with the insect. When it appears, perching on Merwen's shoulder on Valedon, it clicks and squeals, scraping "lopsided" mandibles together, and the pattern Merwen weaves into her seasilk takes on swathes of color "more fantastic than the robes of Iridian nobles" (20). These greens, golds, and blues hint at a sublime mystery in the ecology of companion species on Shora. They are engaged in a hybridizing relationship that makes them amborg subjects.

The clickflies enact a carnivalesque uncrowning that deflates the harm implied by the horrified reaction of Spinel and the other Valans toward them. The sense of threat the Valans feel is out of proportion to the clickflies' appearance and behavior. Growing familiarity with these creatures undermines such fearful responses and illustrates a co-shaping of bodies and histories that is cognizant of animal others as subjects. Part of Spinel's acclimatization and sense of feeling at home is communicated by his growing ability, not just to react, but to respond to the clickflies. Spinel's superior view of the organic and the bodily is uncrowned. Additionally, as he learns to decode the clickfly webs he becomes part of Sharer life, and his orientation to other lives undergoes a transformation.

Toward the end of the story, the Valans attempt to eradicate the clickflies once they discover that they form the backbone of the Sharer communication system. Such a move illustrates the lack of respect the Valans have for animal life. They conceptualize the clickflies as pests and components of a communication infrastructure, not as fellow amborg subjects. The reader, however, is attuned by now both to the physical harmlessness of the clickflies and their importance to Sharer culture, which in turn highlights the extremity and cruelty of the solution the Valans adopt. This uncrowning works in tandem with a range of others, from the "animal" nakedness of the Sharers to their language. For example, when on Shora the commoner Spinel and the noblewoman Berenice adopt Sharer nudity, an uncrowning when their different statuses as imperial subjects and the Valan view of Sharers as primitives is taken into account. Their shared nakedness contributes to the levelling of their status, a transition from commoner and lady to "sister" Sharers.

The clickfly-Sharer relationship exemplifies what Vint (2010), following Derrida, identifies as the basis for the Sharer language: the capacity "to respond, not merely react" (448). This reciprocity is built into the Sharer language: Slonczewski explains that the language is not structured with a subject-object relationship governing agency. Verbs such as "wordsharing" are thus the equivalent of both "speak" and "listen" in the Sharer tongue. Merwen asks a frustrated Spinel, "What use is the one without the other?" (36). The Sharer language underscores a fluidity of experience that Slonczewski links to "the doctrine of 'satyagraha,' love-force, developed by Gandhi and other historical practitioners of nonviolent resistance" (Slonczewski 2001).

The Purple Plague: Symbiogenesis and the Grotesque Interval

The second context for the appearance of the word "grotesque" is when Spinel finds himself succumbing to the purple plague. His cohabitation with the Sharers and their bodies—adapted for existence on the ocean planet—sparks a reaction that had been suppressed until he begins to develop the Sharer's characteristic purple hue. The shock of this misrecognition of his own body transforms the Sharers in his eyes, and he suddenly sees them in a negative way as they "reached out to him with livid limbs and flippers, grotesque signs of what he would become" (97). He escapes into the tunnels of the raft, obliterating in these dim spaces the sight of a skin that horrifies him: "as if by sheer force of will he could keep himself from becoming a monster" (97). This space, between what was and what is becoming, evokes Harpham's notion of the grotesque interval, which Csicsery-Ronay (2008) tells us is that "between the transmutative fluidity of the object and the classificatory uncertainty of the perceiver" (187–88). The cause of the purple plague is the breathmicrobe, a harmless organism modified by the Sharers to increase their oxygen efficiency. This symbiotic microorganism is an instance of a pantropic strategy that enables the habitation of Shora.

Breathmicrobes are apt agents for the grotesque. According to Csicsery-Ronay (2008), its aesthetic is concerned with the smallest of scales, with "cells, genes, molecules, atoms, bytes" (212). Breathmicrobes signify the grotesque porousness of animal bodies; infection by this innocuous microorganism represents a threshold beyond which individuals become enmeshed and begin to flourish within the interspecies communities that make up the planet's ecology. The Valans fear this transformation and take drugs to prevent becoming hosts to the breathmicrobes, refusing on this fundamental level the amborg relationship. The Valans' lack of knowledge about the transformation, and their projected fear that the Sharers will tailor a deadly strain of the breathmicrobes for use as a bioweapon, speaks to the grotesque interval, a contact zone that invites the possibility of a hybridizing response with the other. The Valans misunderstand and simply react to the Sharers because they attempt to fit Sharers into their system of classification. The Sharers' challenge to respond disrupts the classificatory systems that the Valans use to support and justify their colonial endeavor.

Spinel's call to respond comes at this moment of crisis, and he retreats into a "shelter, a cocoon to hide away from" the Sharers (97). Csicsery-Ronay (2008) notes that "The grotesque … turns the arrested attention intensely toward things, in which it detects a constant metamorphic flux, an intimate roiling of living processes that perpetually change before understanding can stabilize them. This process is one of steady descent into interiors, into grottoes of being, in the hope of finding a core, but always finding more transformation" (190). His purple skin, however, is as inescapable as the breathmicrobes that inhabit his body and environment. He is forced to confront the full significance of what it means to be implicated in the shifting processes of a multispecies community. Spinel's descent into the womblike interior of the raft is a descent into what Bakhtin would call the material lower bodily stratum, a term he uses to refer to a series of grotesque bodily images—gustatory, excremental and reproductive—that characterize the carnivalesque uncrowning (Bakhtin 1984, 21). These images create reversals by undermining what was previously valorized: For Spinel, the transformation of his skin color undermines his identification as Valan and destabilizes the boundary separating Valans and Sharers.

Csicsery-Ronay (2008) acknowledges that the grotesque "may appear reductive and essentializing" because it aligns the transformative space with the womb and thus with the feminine, but he also points out that this characterization hints at "a mythological charge that Harpham detects in the grotesque" (194). The mythological and literary coupling of the feminine with such spaces does not necessarily assume an essentialism that sees this association as gender determined, but rather points toward a literary-historical link between the grotesque body and the transformative space of the mythic womb. The use of the grotesque in this ecofeminist novel draws on the association of bodily permeability and womblike spaces with the feminine, although such images are complicated by the lifeshaping that allows all children to be born female.

The raft's outward appearance, inverted trees whose branches plunge the ocean, complements the image of the grotesque space of the tunnels. The grotesque "is looking for that which protrudes from the body, all that seeks to go beyond the body's confines. Special attention is given to the shoots and branches, to all that prolongs the body and links it to other bodies or to the world outside" (Bakhtin 1984, 316–17). The raft's "roots" connect the Sharer habitations to the surrounding seas, and they provide shared living space for other animals that live as companions with the Sharers. The Sharers' biomedical technology, part of the raft system itself, involves the use of vines that enter or, as Spinel characterizes it, "worm" into the body to make various adjustments to its functioning (92). The medical and the grotesque are coupled, as are symbiogenetic transformations through the purple plague and the transformative spaces of the raft. These transformations are examples of the pantropic themes central to *A Door into Ocean*.

Spinel's descent into the raft's life-generating depths is a confrontation with the grotesque body of the self as other, a transformation that turns the gaze inward. His struggle involves a re-categorization of the boundaries between self and other, a transformative crisis that results in his accepting and responding to the multispecies community of the Sharer raft. Spinel, as a Valan and newcomer to Shora, begins to adapt physically and experientially to habitation on this new world: He undergoes pantropy in his own person, both via colonization by the breathmicrobes and by intermeshing with wider multispecies communities in a symbiotic ecology. Spinel's reaction to his transformation, to the emergence of signs of the animal within his own body, contrasts with that of the Valans, who are later colonized by the breathmicrobes. In one dramatic scene in which five Sharers protesting outside a Valan base enter whitetrance (reducing their bodily functions to the bare minimum for survival), the Valan commander Realgar sees "five pallid Sharers facing soldiers awash in dreadful violet, as if the color had seeped directly from one side to the other. No one spoke, except the inescapable voice of the sea" (299). The Valan sense of superiority, marked by a difference in skin color and clothing, is subjected to an uncrowning. Many Valans react poorly to this instance of unintentional pantropy, and notably they plan to terraform the planet rather than accept adaptations to their bodies or to their mode of living on Shora.

Humans and the Anthropological Machine

The Valans' orientation to the Sharers involves identifying them as animal even when knowledge of their shared evolutionary origin becomes widespread. This resistance to scientific taxonomy speaks of another system of classification that depends on the hierarchical ordering of groups according to their behavior, rather than their genetic similarity or concordance:

> "Well, now. You have to understand, sir, when I first came out here twenty years ago, the catfish weren't considered human at all, just another part of the natural fauna. You can't even mate with them properly, and to my mind"—Kyril shifted his weight and sat forward. "Look, if you're going to call them people, you better understand just what kind of people they are. They never had to back down to anyone, not since before the rise of Torr, and they just plain don't know how. You'll have to blow up half the planet to teach them." (243)

The Valans' colonial folk taxonomy centers on dominance as the basic structure for categorization. Attitudes and practices that cannot be accounted for by this system pose problems for how the Valans should order their relationships with both the people and other animals of Shora. By doing so, they also exclude the attitudes and behavior that characterize the Sharers' pacifism and investment in multispecies relationships. As Kyril the merchant struggles with the evidence of their shared origins, his compromise entertains the possibility of classifying them as people only insofar as it makes scientific sense to do so. It is clear that for Kyril, despite the incontrovertible evidence of their humanity according to evolutionary biology, Sharers still remain "catfish" because they do not share Valan values. A difference in political and ethical systems forms the basis for a division between the human and the non-human. Like their colonial counterparts, scientific evidence fails to convince the Sharers to reframe their view of the Valans, since the Valans do not act according to values the Sharers recognize as human.

Merwen's insistence on the mind as a center for the definition of the human is a typical move that reflects some aspects of human exceptionalism. When Merwen asks Spinel whether a monkey that she sees on Valedon is human, Spinel replies, "are you kidding? ... It's a monkey. People eat them, even." Her retort, "How was I to know? You're human," speaks to the grotesque indeterminacy that lies embedded within the border between humans and monkeys and which troubles these neat classifications (39). Merwen's failure to recognize the monkey as other illustrates one point of departure from Valan taxonomies. Merwen's recognition of herself in the body of the monkey, and Spinel's failure to do so, highlight two different approaches to the question of what it means to be human.

Unlike some of the other Sharers, Merwen is convinced of the humanity of the Valans, and her adoption of Spinel is in part an attempt to prove this fact to her community. Speaking of the ability to recognize oneself in the other, and the other in the self, Merwen tells Realgar that "there is a difference between seaswallower and human. A human sees herself in the mirror. I am human, and so, inescapably, are you" (367). Realgar's attempt to subvert Merwen's pacifism by having her accept his logic of extermination in the face of Sharer-Valan conflict is designed to overcome any notion of similarity between Merwen and himself by, paradoxically, having her admit that her approach is fundamentally untenable and that the Valan approach—violence—is the only reasonable response. Such a concession would give Realgar the justification for ordering the genocide of the Sharers because it would be reasonable to assume that if the Sharers accepted his logic of violence, it would be permissible for them to use their advanced lifeshaping skills for the purpose of devising biological weapons.

Merwen's response is to emphasize the human ability for self-recognition, but undergirding this basic notion is another category fundamental to Sharer philosophy: "A lesser creature sees its rival on the water and jumps in to fight it. A human sees herself and knows that the sea names her. But a selfnamer sees every human that ever was or will be, and every form of life there is. By naming herself, she becomes a 'protector' of Shora" (61). This three-tiered taxonomy introduces another category to the already troubled human–animal binary. While Merwen's description of the "lesser creature" draws from studies of the animal's failure to recognize its reflection, it also chimes with the Valan approach to the Sharers; they fail to see themselves reflected when they gaze at their counterparts. They are constructed—not as animal—but as subordinate to the other two categories. This distinction accounts for the Sharers' inability to accept Valan humanity, given the Valan record of violence on the planet. Humans, on the other hand, are able to recognize themselves in the image that the ocean reflects back at them. To be a "selfnamer," however, requires something more. It necessitates an understanding and acceptance of the self as embedded in multispecies relationships. It requires a "long view," one that can accommodate the temporal and spatial relationships between humans and other animals, along with their interconnectedness across time and space. This view aligns the selfnamer with sustainability as a mode of consciousness and to the recognition that the amborg subject is constituted by individuals in relationships of difference and similarity. By accepting such a mode of consciousness, the selfnamer is better able to protect Shora, companions, and future generations.

Seaswallowers and the Grotesque Body of the World

Seaswallowers are cephaloglobinoids, creatures presumably descended from squids, with life cycles that include tempestuous breeding periods where they agitate the seas with whirlpools recalling Charybdis. Csicsery-Ronay (2008) points to this Homeric beast as a source for the grotesque trait of containing two bodies in one: "a new body may be in the process of metamorphosing out of an old one; a being may combine, conflate, or be trapped in two corporeal forms; a body's appearance may conceal a completely different one underneath" (197). Companion species presumes this ongoing transformation involving bodily reversals, blending, and emergence. "The grotesque body … is a body in the act of becoming. It is never finished, never completed; it is continually built, created, and builds and creates another body. Moreover, the body swallows the world and is itself swallowed by the world" (Bakhtin 1984, 317).

The seaswallowers exist as part of the grotesque transformations of Shora. Despite the Sharers' extensive knowledge of biology, down to the cellular and atomic level, knowledge of the seaswallowers is scant. They are re-formulations of the sandworms of *Dune*; "beasts of the deep, they swallow all in their path" (16). Like the sandworms, they are symbols of nature that combine death and life into one body. The destruction that they bring is essential, not only for their own existence, but also for the flourishing of others: "seaswallowers passed as they always did, leaving the waters clean and clear" (396). Without their bi-annual clearing of the waters, the Sharers' way of life and the lives of their companion species is jeopardized. Bakhtin suggests that "the combination of killing and birth is characteristic of the grotesque concept of the body and bodily life" (248). As the Valans destroy the seaswallowers to minimize the risk of damage to their property, the Shoran ecology is diminished. "Without seaswallowers … the entire life web would collapse, and Sharers would starve" (367). They are keystone species, ambivalent companion animals who make Sharer life possible.

The seaswallowers' consumption, in this ecological context, is also a regeneration of the cosmic body. It is an essentially generative image, one that capitalizes on a series of images related to the material lower bodily stratum. These images are an essential key for understanding the grotesque in *A Door into Ocean* as an ecological aesthetic that links the image of Earth to that of the individual. Bakhtin contends that "death, the dead body, blood as a seed buried in the earth, rising for another life—this is one of the oldest and most widespread themes. A variant is death inseminating mother earth and making her bear fruit once more" (1984, 327). This image is connected to terraforming, which requires the death of microbes and plant-life to bring another kind of life to planets.[2] These are generative images that turn inward on themselves, embodying the essential ecological reversal of a transformation from death to life at a macrocosmic scale.

This connection between the individual and the wider environment is coded into the reciprocity of the Sharer language. When attempting to explain to Spinel the variety of verbal "share-forms," Merwen confirms the validity of what Spinel believes are examples of the impossibility of jettisoning the object-subject distinction. One exchange in particular is especially telling:

> Or if I swim in the sea, does the sea swim in me?
> Does it not? (37)

This flow between interior and exterior, the individual and cosmic body, connects companion species and the grotesque to issues of landscaping embodied by the terraforming motif. It becomes possible to read terraforming as a way to intervene and transform aspects of the individual body and the communities in which they are embedded. The image of the ocean planet emphasizes this flow between organism and world and, with the breathmicrobes factored into this reciprocity, it connects the grotesque directly to the permeability of multiple bodies in multispecies community. In a very literal way, the sea does permeate the submerged body; microorganisms traffic between world and organism and confuse the supposed distinction between autonomous bodies.

This traffic between boundaries is encapsulated in another image that takes the grotesque aesthetic and merges it with the sublime. As the seaswallowers pass, the second cycle of regeneration begins, with bioluminescent protozoans able to multiply unchecked in the wake of the devastation the seaswallowers cause: "in a few days water-fire bloomed again, a lovely bioluminescence that etched the waves for seven nights and kept everyone awake dancing in its brilliance. Another strand of the living web was rejoined" (396). The grotesque and the sublime are entwined in this image. Csicsery-Ronay (2008) notes that "it is not always easy to distinguish the two modes, as they are dynamically, dialectically related. A phenomenon that to one mind appears to be grotesque may appear sublime to another, if the principles behind it are seen not as violations of reason, but its primal processes" (146–47). To the Valans, the seaswallowers and the Sharers' adaptation to their life cycle are grotesque because they violate the fundamentally hierarchical relationships that humans should build with other animals within their philosophy. Yet understanding the place of the seaswallowers as one of many companion species highlights their fundamental role in renewing the cosmic body of the world. As such, the destruction and renewal that follows can be seen as a sublime recognition of necessary ecological relationships, encoded by the appearance of firewater. As another example of the traffic of microorganisms between bodies, the Sharers ingest the diatoms that make up firewater: "it was a spectacular time for night dipping, to bathe oneself all over in the water-fire, to swallow the 'flames' and spit them out again. One's teeth glowed for hours afterward" (161). The Sharers revel in what the Valans interpret as grotesque disturbances between bodies.

In conclusion, I have considered the significance of the grotesque in the multispecies communities depicted in *A Door into Ocean* with the aim of illustrating how this aesthetic form can be incorporated into a biopolitical framework for animal studies in sf. I have extended Vint's analysis of biopolitics and the human–animal relationship to provide a foundation for continuing research into the intersection between terraforming, pantropy, the animal, and the grotesque. I aim to rehabilitate the category of the grotesque as a powerful aesthetic for exploring the human orientation to nature as a whole, and to animals in particular. The grotesque offers a foundation for considering taxonomic classification and the multiple ways in which different communities order their relationships with other bodies and groups of bodies, including the cosmic body of the world. The grotesque "seeks to grasp in its imagery the very act of becoming and growth, the eternal incomplete unfinished nature of being. Its images present simultaneously the two poles of becoming: that which is receding and dying, and that which is being born; they show two bodies in one, the budding and the division of the living cell" (Bakhtin 1984, 52).

These continuous transformations require flexibility when dealing with relationships between the human and non-human and call for a promiscuous realism that can adequately take cognizance of the different and sometimes conflicting ways in which humankind relates to animals. The Valans' colonial endeavor, predicated on attempts to reduce species multiplicity by erecting barriers between themselves and others they regard as animals, requires the subjection of others to their desire and control. *A Door into Ocean* explores what it means to be an amborg subject made up of individuals whose relationships are predicated on both response and respect. In contrast to the extermination of life that the Valan terraformation of Shora would bring, *A Door into Ocean* offers a vision of pantropic relationships based on the evolution of symbiotic relationships that work to create environments, and ultimately worlds, within which multispecies communities can flourish. This relationship is itself an alternative form of terraforming based upon the elementary interrelatedness of life and death and the embeddedness of grotesque bodies within the cosmic body of the world.

Notes

1. See Pak (2010) on vivisection in *The Island of Dr Moreau*.
2. See Pak (2014) on the emergence of life from death in the context of terraforming.

References

Agamben, Giorgio. 2004. *The Open: Man and Animal*. Stanford: Stanford University Press.

Bakhtin, M.M. 1984. *Rabelais and His World*. Bloomington: Indiana University Press.

Blish, James. 1942. Sunken Universe. *Super Science Stories*, May, 49–63.

———. 1952. Surface Tension. *Galaxy Science Fiction*, August, 4–40.

———. 1956. A Time to Survive. *The Magazine of Fantasy and Science Fiction*, February, 15–60.

———. 2001 [1957]. *The Seedling Stars*. London: Gollancz.

Csicsery-Ronay, Istvan, Jr. 2008. *The Seven Beauties of Science Fiction*. Middletown: Wesleyan University Press.

Dupré, John. 2002. *Humans and Other Animals*. Oxford: Clarendon Press.

Gordon, Joan. 2008. Gazing Across the Abyss: The Amborg Gaze in Sheri S. Tepper's "Six Moon Dance." *Science Fiction Studies* 35 (2): 189–206.

Haraway, Donna. 2008. *When Species Meet*. Minneapolis: University of Minnesota Press.

———. 2013. Sowing Worlds: A Seed Bag for Terraforming with Earth Others. In *Beyond the Cyborg: Adventures with Donna Haraway*, ed. Margret Grebowicz and Helen Merrick, 137–46. New York: Columbia University Press.

Herbert, Frank. 1965. *Dune*. Philadelphia: Chilton Books.

Le Guin, U.K. 1976. *The Word for World Is Forest*. New York: Berkley Books.

———. 1996. "The Carrier Bag Theory of Fiction." In *The Ecocriticism Reader: Landmarks in Literary Ecology*, ed. Cheryll Glotfelty and Harold Fromm, 149–54. Athens: University of Georgia Press.

Pak, Chris. 2010. The Dialogic Science Fiction Megatext: Vivisection in H.G. Wells' *The Island of Dr Moreau* and Genetic Engineering in Gene Wolfe's "The Woman Who Loved the Centaur Pholus". *Green Letters* 12: 27–35.

———. 2014. "All Energy Is Borrowed"—Terraforming: A Master Motif for Physical and Cultural Re(Up)Cycling in Kim Stanley Robinson's *Mars* Trilogy. *Green Letters* 18 (1): 91–103.

Slonczewski, Joan. 1986. *A Door into Ocean*. New York: Tor Books.

———. 2001. *A Door into Ocean: Study Guide*. http://biology.kenyon.edu/slonc/books/adoor_art/adoor_study.htm. Accessed 26 Sept 2019.

Vint, Sherryl. 2010. Animal Studies in the Era of Biopower. *Science Fiction Studies* 37 (3 November): 444–55.

Williamson, Jack. 2004. Collision Orbit. In *Seventy-Five: The Diamond Anniversary of a Science Fiction Pioneer*, ed. Stephen Haffner and Richard A. Hauptmann, 216–77. Royal Oak, MI: Haffner Press.

CHAPTER 5

Bodies That Remember: History and Age in *The Children Star* and *Brain Plague*

Derek J. Thiess

The act of creating a history is a dynamic process of writing and revision laden with ethical pitfalls. As Hughes-Warrington (2013) recently suggested, history requires ethical attention to the fact that "revision in history is … in one sense dispersion history. Writers not only narrate the dissipation of a particular view of the past; they also in some cases seek it" (32). *The Children Star* (1998) begins in a desolate landscape plagued by various histories under revision. The planet L'li is dying—overpopulation has led not only to abject poverty but to disease and ruin—and Brother Rod, a Spirit Caller, has come on a mission searching for survivors to populate a new world. When Brother Rod finds an abandoned child, 'jum, his intent is to take her away from L'li to a new planet, Prokaryon, nicknamed "the Children Star" because it has not been terraformed, thus settlers must be "lifeshaped" to adapt to its climate, a process much easier in children. However, the desolation of L'li is not random. Elysians populated the planet and allowed it to expand in order, in part, to increase their wealth and dominance in the Free Fold. More importantly, unlike

D. J. Thiess (✉)
Department of English, University of North Georgia, Watkinsville, Georgia

© The Author(s) 2020

B. Clarke (ed.), *Posthuman Biopolitics*,
Palgrave Studies in Science and Popular Culture,
https://doi.org/10.1007/978-3-030-36486-1_5

the Children Star, L'li is a terraformed world, boiled off and stripped of its original life in order to be made suitable for human occupation. Ecocidal planet cleansing is a fitting analogy for the action of historical revision, particularly in approaches to history that, however well-intentioned, seek to displace one history for another. Moreover, the novel's own language supports an analogy between terraforming and historical displacement, when its Elysian antagonist Nibur Letheshon suggests that planet cleansing is "Like taking down a historic building—once the first corner goes, no more protest" (161). Like the productive destruction of a planet or a building, revisionist histories are ultimately about forgetting, about the destruction of another, prior history.

In the same passage in which Nibur connects historical destruction with physical space, he also expresses his annoyance at the delay of his ship because of the outbreak of a contagion: "Mortals and their sickness—they should manage their bodies better" (161). For Nibur, the bodies of mortal humans who age naturally, unlike those of the immortal Elysians who are engineered against both age and contagion, are subject to the same processes of revision.[1] In Slonczewski's ecofeminist Elysium novels, matters of embodiment highlight the displacements of the history that is to be rewritten by the powerful. Moreover, in paying special attention to bodies for which a range of ages is important, this novel in particular can be read as drawing attention to the shortcomings of cultural theorizations of embodiment that exclude age in discussions of intersectional gender, race, sex, and orientation.[2] Discourses of power almost always exclude discussions of the differently aged. Like the ageless Elysians, some social-constructivist criticism regards biological difference in age as outside discourses about equality. However, *The Children Star* resists this process by including the physical and biological effects of aging in a social critique of power structures. My reading connects these thematic issues of aging bodies to critical debates over historical revisionism.

Both the freedom to recreate history as one pleases and a tendency to overlook the biological materiality of aging are symptoms of the recent privileging of social-constructivist approaches to history and science, respectively. And there is good reason for this in the tendency of historical work to universalize a white, European, male perspective. Slonczewski's novels offer a different biological narrative in which naturally aging bodies and their marginalized positions call attention to the biological limitations of the human. They remind us of the materiality of history by reminding us of our place in the biological kingdom. In this sense, these bodies foreshadow those microbial bodies of the "hidden masters" of *The Children Star* and *Brain Plague*, revealing that "humans and animals alike are shaped and controlled by modes of biopower that designate ways of living and dying" (Vint 2010, 444). This is an unwelcome reminder of our mortality, and herein one may see the intrinsic connection between the materiality of historical destruction and our wish to avoid the biology of aging.

The writing of history tends to universalize a white, European, male perspective. In contrast, the Elysium Cycle presents a biological narrative in which naturally aging bodies and their marginalized positions call attention to the biological limitations of the human. *Brain Plague* in particular demonstrates even more clearly that there are limitations and consequences to this forgetting. In this story, the lines between physical place, discourse, and biological body disappear. Attebery (2002) has declared that "there are no bodies in fiction, only words that call bodies to mind" (16). His point is valid, and Attebery sees this as an advantage upon which SF may seize to explore how gender is coded and to generate new potentials. However, the last book of the Elysium Cycle will not allow us such an easy separation of word, body, and history. As the sequel to *The Children Star*, the microbial life discovered there has spread to other worlds, both intentionally and accidentally. Chrys, an artist struggling to make a name for herself, volunteers to host a colony of "micros" within her body, hoping to gain a competitive advantage from their mind-enhancing potential. She soon discovers that she is no longer alone—not only does her society of micros keep her constantly occupied, communicating via their home at the base of her brain stem, but they also continually interact with other micro-societies in other hosts. Moreover, the micro-society inside Chrys is known for several unique traits: They are architects that have created some of the most impressive sentient buildings on Iridium, they are decadent libertines, and they are meticulous historians.

Unlike the bioengineered immortal Nibur, however, micros *must* value human bodies because such bodies are literally their planets. Bollinger (2010) characterizes Chrys's relationship with the micros in terms of symbiogenesis, in which the functional merger of separate organisms "generates a new self, amplified and improved but still with full agency for both participants" (47). For Bollinger, this signals a positive posthuman cooperation between human and micro. One must also note that Chrys's situation is an embodied position—bodily boundaries can be redrawn, but the biological substance of the body retains an important position in the social and historical lives of both human and micro. Chrys's micros build their homes with Chrys's own proteins, encoding their history into the walls of those homes. In stark contrast to the "ageless" Elysians, their life cycles are so short by human standards that they place great importance on memory, and they must constantly pass their history on to their own children. In this scenario, to destroy the human body is to destroy one's planet is to destroy the inscription of history. *Brain Plague* makes evident the relationships among the materiality of history, its grounding in place and person, and the biological constriction of age upon the body. With their accelerated lives, the micros demand that we not forget our place in that cycle of living and dying.

SF and History

looseness-2SF criticism has generally positioned itself against a notion of physical consequence or responsibility in relation to actual history. For example, Csicsery-Ronay (2008) has cataloged the various ways in which SF engages "future history." Reflecting a trend away from modernity in scientific and historical work outside of SF, future histories have turned from the evolutionary chronotope identified with determinism in the work of writers such as H.G. Wells to engage instead what he calls (following John Clute) "History Free and Experimental." This has happened because "SF lacks the gravity of history, because it lacks the gravity of lived experience. It is *weightless.* Its represented futures incur no obligations... . It may be persuasive, but it is always poetic and figural" (2008, 83). This freedom to explore time and space is a cornerstone of SF literature, and certainly, the Elysium Cycle participates in this tradition. Whether in retrofuturist novels, alternative histories, or other modes, such feminist SF destabilizes established narratives of history in highly productive ways. As Wolmark (2005) has succinctly summarized,

> An open-ended sense of identity, and the capacity for it to be constructed and reconstructed in time and history, allows for the creative destabilization of definitions of self and other and for the acknowledgment of difference. The contested futures of feminist SF provide the arena in which the disruptive consequences of such destabilizations can be explored. (161)

The worlds of feminist SF, in particular feminist utopian societies such as that of the Sharers in *A Door into Ocean*, are created via the disruption of established narratives of history and embody these destabilizations in their politically and sexually open and free social orders.

But the notion of utopia complicates matters—and it is certainly possible to cast as utopian the desire to exist "*to the side* of consensus reality ... without the sense of urgency and implication that comes when the future is felt to be at stake" (Csicsery-Ronay 2008, 103). Utopian desires complicate matters because of the impossibility of entirely avoiding responsibility for creating the world via our choices and needs. Gomel (2014) advocates a return to the twentieth-century utopia as a means to "understand the ever-present danger of trying to impose a definitive shape upon history, to contain the energies of time by imprisoning it in space" (121). She suggests a kind of "free history," a postmodern history in which the "'death of history' is projected upon impossible spaces, in which distance and difference are abolished, and past and present coexist" (37). But Gomel's is a guarded suggestion, because even within these impossible spaces there exist "zones of temporality" that become heterotopias and heterochronies. Moreover, to Gomel, these heterotopias are utopian filters—they are the state of exception that must exist in order to permit utopian sovereignty. The epitome of the heterotopias in the various chronotopes that Gomel examines is, as in Agamben's theory, the death camp. In this formulation of SF time and space, it is impossible to avoid fully the responsibility to "lived experience" that historical work occasions.

The persistent presence of the heterotopias makes it necessary to accept the ambiguity of both this desire for freedom and its impossibility. For example, even the Omelasians' great joy because of the child who suffers in the basement is not a "vapid, irresponsible happiness" (Le Guin 1973, 283). Le Guin's child, whose youth as compared to the age of those who benefit from its suffering, reminds us again of the connections between history, age, and time. *The Children Star* and *Brain Plague* are fitting sites to which to bring these considerations. We will follow two narratives that converge in these works: One involves the revisionist process of spatial-historical destruction and the other displaces onto naturally aging bodies our own fears of the biological. In one narrative, a planet may be destroyed so that humans can create a new world for themselves. In the other, the differently aged may be erased so that "adult" humans may enjoy "full life" and continue their search for meaningful immortality: "In remembering 'something else,'" such relativistic approaches to history "also forget, in a very concerted way, a particular history that does not accord with the manner in which they want to view the past" (Thiess 2014, 162). This impulse to forget applies equally to our biology, and considering these two narratives in concert suggests the need for biopolitical criticism to include an attention to aged bodies. If, on the one hand, "there is no 'the body': there are only various bodies differentiated by endless permutations of race, class, age, gender, sexual orientation, geographical location, and any other category we use to discipline and value bodies" (Vint 2007, 184), on the other hand, feminist gerontologists have also suggested that the categories of race and gender in particular have received a great deal more attention than the category of age. Placing the physical aspects of aging at the center of narrative discourses about spatial-historical destructions, *The Children Star* and *Brain Plague* demonstrate that this displacement of age is symptomatic of our fears of a very personal history of biological limitations, of living and dying.

Destroying Place-History

In *The Children Star* (1998), L'li and Prokaryon are both colonized planets. What differentiates the two worlds is that, unlike L'li, "Prokaryon was not terraformed, for today the Fold forbade alien ecocide. But Prokaryon's alien ecology, full of arsenic and triplex DNA, poisoned human bodies. Unless of course they were lifeshaped, their genes modified to survive" (15). Even this distinction may be tenuous, however, as L'li is currently being ravaged by a deadly prion disease. Brother Rod distinguishes the dangers of the two worlds, one where "prions arose from human bodies" and another "where humans could barely live" (24). In this juxtaposition, planetary ecosystem and human body continually call one another into question—their living and dying mirror one another. Yet in these instances, the relationship between planet and body is not as symbiotic as elsewhere in the Elysium Cycle, for Prokaryon's ecology is toxic to the human body and the presence of humans on Prokaryon threatens its ecology.

The novels of the Elysium Cycle are all about co-existence, just as Slonczewski, as a working biologist, emphasizes the importance of symbiosis. *A Door into Ocean* portrays the difficulties that arise between the Sharers, the Valans, and the Patriarchy of Torr. *Daughter of Elysium* depicts a similar struggle, as humans are required to acknowledge the coeval sentience of their intelligent machines. *The Children Star* continues the imperialist encounters of the prior novels while differing from them in that the settlers to Prokaryon are not changing the planet to suit their needs. Life on Prokaryon is based on triplex DNA, making impossible the somewhat simpler (biological) co-existence of the previous novels. Thus, the children brought there must be lifeshaped; however, many feel this burden is unsustainable, and so characters such as Nibur seek to terraform the planet to allow for its more immediate colonization. The sentient microbes, the "hidden masters," however, halt this process, because the law forbids their destruction once located *and* because they are able to lifeshape humans to live on Prokaryon more cheaply and quickly than the human process. This is a different kind of symbiosis in that human inhabitants and their adopted ecosystem now adapt to one another via an intermediary, much like the helpful role of the microbiota that already exists in the human body. Acknowledging the reality of this new symbiosis will be the challenge resolved within the storyworld of *The Children Star*.

In fact, in Brother Rod's distinction between the dangers of a world of prion disease and a world with an arsenic-based biology, one finds the first hint of the tension that will occupy the novel. Rod muses further on a "release form [that] required all immigrants to acknowledge that Prokaryon's biosphere was only partly understood, and its climate not yet controlled, and that the appearance of any plague threatening the Fold might require defensive action—before all inhabitants could be evacuated" (1998, 23–24). The characters in the novel are acutely aware that this "defensive action" could well mean the terraforming of the planet, the erasing of its history—and it is no accident that it is a written document, a discursive form, that allows for this historical destruction. The only thing that has kept the Free Fold from terraforming Prokaryon thus far is the search for sentient life. Rumors persist that "There are intelligent aliens running Prokaryon... Can you explain how Prokaryon has all those rows of forest, one after another, all across the continent? Who tends the garden?" (20). According to the laws of the Free Fold, if such sentient life is discovered and deemed "human," the planet may not be terraformed. This search for intelligent life on Prokaryon occupies the Spirit Colony in this novel as they race to try to save their home.

The immortal Elysian, Nibur, has already set in motion plans to first purchase control of the planet and then to cleanse it, to tear the historic building down. Nibur justifies his actions historically: "clearing land is something humans have done since they first evolved" (108). Nibur seizes the opportunity to "settle millions of starving L'liites on a free world. Humans with vital needs—and souls remember," contending with regard to the suspicion of sentient life on Prokaryon, "we Elysians needn't worry to keep body and soul together. How dare we put the needs of some mythical creature above our own human kind?" (119). His comments also juxtapose the human body with the planetary surface. Even more specifically, the reference to "mythical creatures" calls to mind legendary history and recreates a long tradition of "mythic" histories being supplanted by modernization.[3] The intent is literally to deface the planet's surface and give it a new one, like melting a wax tablet or writing something new on a page. The relationship between physical, historical place or structure and the writing of history is a long and tense one. In many ways, the philosophy of history has for the last few centuries meant to "challenge the idea of histories as stable and bounded places" (Hughes-Warrington 2013, 50). Yet critics such as Hughes-Warrington (2013) suggest that this challenge, too, may be a kind of escape: "it is because it is possible to go and relax in a café that one tolerates the horrors and injustices of the world without a soul. The world is a game from which everyone can pull out and exist only for himself, a place of forgetfulness" (59). The challenge to place-history potentially makes the creation of an alternate history an act of forgetfulness that may simply efface one space to create another, planetary or otherwise. Nibur is a historical relativist along with the critic Southgate (2013), in whose words, although we cannot "jettison the past and forget it," we "can surely encourage preferable dreams" (200).

Nibur does seek to "jettison the past and forget it"—to destroy the surface of the planet and inscribe a new history onto its surface. This is an approach to history I call "lethetic reading," a mode of reading that "allows the reader to 'disbelieve' anything… . It need not be a conscious disbelief, but may also be simply 'forgetfulness,' a mere byproduct of focusing on the word and letting the world slip away" (Thiess 2015, 41). In the story this novel tells, the idea of letting a world slip away acquires new meaning as Prokaryon's history faces immediate destruction. Moreover, Nibur Letheshon is well-named after Lethe, the mythical river of forgetfulness. Another example of forced forgetfulness is the way that Nibur's giant starship, Proteus, must have its networks cleansed every few hours in order to halt the process of its imminent sentience. In the Elysium Cycle, complex machines can evolve to sentience, and those that do achieve independence within the Free Fold. However, Nibur has developed a method to halt sentience, including this continual "cleansing" of networks, onto which activity he "had steadily built one of the largest commercial empires in the Fold" (112). The spatial-historical metaphor is evident, as is the forgetfulness implied by revision—Nibur builds a new edifice upon the ruins of those he has already cleansed and forgotten; he lets one world slip away to create another.

There is a tension, therefore, between the "ageless" Elysians, who function as organs of forgetting, and the other races of the novel, who take great care to remember their much shorter lives. In fact, the Elysians have grown decadent in their near-immortality, modifying mortal, biological, and planetary histories for aesthetic reasons alone. Thus, Nibur's concerns in *The Children Star* do not focus on ethical considerations but rather on how he might remake the interior of the Proteus into a replica of Prokaryon's Spirilla continent. He deems this an "esthetic challenge" (21). As an architect working with nanoplast—intelligent building material composed of nanomachines—Nibur may "immortalize" the places he destroys, like collections in a giant *wunderkämmer*. While in *The Children Star* this behavior seems coolly decadent, however, there are suggestions in the other Elysium novels that the aestheticization of these places betrays a kind of psychological response to the Elysians' extended longevity. One Elysian character in *Daughter of Elysium* wonders, "How solitary we are ... each one of us with our layered histories floating in a sea of time" (72). While one might read this notion of layered Elysian history as a palimpsest of sorts, it is also possible to read it within the context of a cyclical spatial-historical destruction and reconstruction. One history is destroyed in order to build another, in a repetitive act meant to sublimate the Elysians' loneliness.

The reconstruction of Spirilla on the continually cleansed environs of the Proteus, therefore, is a concerted act of spatial reconstruction for the sake of personal gratification. Hughes-Warrington (2013) makes this connection between historical revision and psychology clear when she suggests that "spatial imagery and language are used to construct temporary textual spaces in which rival historiographies are excluded and readers are encouraged to navigate to a new vision of history" (34). The creator of these histories maintains complete control of this newly captivated environment, just as Nibur controls Proteus with absolute authority. For Hughes-Warrington, however, this control is maintained "in order to protect ourselves from the burden of making decisions in a complex and even overwhelming world" (52). And Hughes-Warrington is not the only critic to suggest psychological motives in acts of revisionist alternate histories.[4] Butter (2009) argues, in reference to the ubiquity of alternate histories about Nazi Germany, that "American culture uses the discourses of uniqueness and good and evil to *displace* and dismember the more problematic aspects of its own history" (109; emphasis added). I have elsewhere suggested that the "Lethetic reading of history … allows a revision of that history to one more conducive to a personal and religious worldview" (Thiess 2014, 161). Viewed from this angle, then, Nibur's recreation is an evasion of ethics in order to maintain his historically destructive, aesthetic worldview, and the responsibility he bears in the destruction of one ecosystem is cleansed by its recreation aboard his ship. Whatever his motive, however, the destruction of a planetary ecosystem must be seen as both a biological and a historical calamity.[5]

Put another way, what Nibur's Proteus affair calls to mind is what Gomel (2014) describes as wormholing, the narrative collapse of time and space upon themselves. Spirilla may virtually exist again on a planet or on a ship orbiting that planet, which status pulls it not only out of space (as place) but also out of time, as Nibur seeks to immortalize it. This connection is more than metonymy—like a wormhole, it represents the folding of both space and time upon themselves in an atopic recursivity. But this atopia dialectically suggests once again those heterotopias that "are not to be seen as a valorized alternative, a sheltering space…. They are often violent, scary, and dangerous. What makes them subversive is their resistance to the totalizing discourse of social perfection embodied in the homogeneity of the text's narrative space" (Gomel 2014, 121). Within the free and experimental history, any zone of temporality—a space in which history is determined or predictable rather than "based on uncertainty and contingency"—becomes a heterotopia (Wolmark 2005, 169). However, *The Children Star* gives this formula an interesting turn as the two spaces—Spirilla on the ship and Spirilla on the planet—are not easily identifiable with either atopia or heterotopia. One would want to associate the frozen, static cross-sectioning of history that Proteus's creations represent with heterotopia, because that would for most critics signal the danger of the determined and determinist approach to history. However, does Nibur (and even potentially Proteus) not have the power to reconstruct Spirilla precisely as he pleases? To continue to shape and reshape it, according to whim or even to his own "preferable dreams"? In fact, to the relativist his recreation of the continent is already shaped by his own subjectivity. The problem this text dramatically portrays is that there are "violent, scary, and dangerous" consequences to this utopian free play for the "real" Spirilla. Historical free play may be especially destructive in its biological effects.

This conclusion is supported by the new kind of symbiosis demanded by Prokaryon's "hidden masters." They are an intervention into our understanding of microbial life because, as the narrator informs us, "even among ordinary pathogens, the most successful eventually mutate to coexist with the host—millions on our skin, and billions in our intestine. What of an intelligent pathogen who *remembers its history* and values its host?" (1998, 238; emphasis added). As the reader finds out later in *Brain Plague*, the micros encode their histories in the proteins out of which they build their cities within the bodies of their hosts. To the micros, biological body and written history are one and the same. Thus, when Rod and his children successfully manage to avert the destruction of Prokaryon, they also save their own physical bodies and the history of the unique culture of micros written onto those bodies. Rod will laugh at the irony when his micros ask him, "HOW DO WE THANK YOU FOR SAVING A WORLD?" (186). The human colonists save Prokaryon, but also and more immediately, they save the world that the micros occupy, that is, Brother Rod's own body. To state this interpretation in reverse, they have saved a biological body that is also a dwelling place and that encompasses the written history of a unique culture. They have reshaped the boundaries among history, place, and body.

This is not to say, however, that the analogy among place, body, and written history always holds. For example, previously the micros tried to thank Rod by directly controlling the pleasure centers of his brain. Earlier yet, they had attempted to control him directly (as they do the nonsentient life forms on Prokaryon) by accessing his pain receptors for a kind of operant conditioning. Immediately after these prior micros declared "ANIMAL WILL LEARN," Rod experiences how "pain filled his skull, spreading throughout his body, as if he were ripped apart very slowly ... it was everywhere, in every limb and crevice, within belly and brain" (165). One might easily read this as a reduction of Rod's body to mere place or thing, a space with distinctly discursive limits. Social approaches to pain have indicated that these limits must be acknowledged: "physical pain—unlike any other state of consciousness—has no referential content. It is not of or for anything. It is precisely because it takes no object that it, more than any other phenomenon, resists objectification in language" (Scarry 1985, 4). The social-constructivist approach to history often emphasizes the limits of discourse in accessing a material past. The point of the present essay is not to negate the limits that discourse places upon history, but to stress the necessary and ethical inclusion of biology as an equally reflexive constraint upon discourse. Despite historical desires to deny our own biology, what we might more ethically acknowledge is the artificial boundary separating biology, place, and history.

What Rod objects to, then, is precisely what Nibur objects to—being considered an "ANIMAL" with a body subject to the constraints of biology, a history situated in a biological place that he would prefer to forget. However, a mode of symbiosis that combines the materiality of history—including our own biological existence—with the discourse of history is certainly possible and definitely preferable. The Elysium Cycle foregrounds biology and in particular the physical body as both place (the world in which the micros live) and history (the site in which the micros write their history and Prokaryon's planetary history). By highlighting the materiality of the destruction of history and making that destruction no less than the destruction of the human body, these novels hold a mirror up to our biology. They force us to confront the fact that we would rather forget the limits of our physical bodies. The most striking of these limits occurs in relation to the aging process in the novels, emphasizing how age is a clear biological constraint and a constant preoccupation of the characters. There is a kind of age normativity in these novels that centers on the "ageless" Elysians and tends toward middle age. As the next two sections show, the latter novels of the cycle involve explicit confrontations between the differently aged—children in *The Children Star* and the aging in *Brain Plague*—and force us to remember our history and our biology, encoded as they both are in our bodies.

Children Who Remember

The problem that will arise for Nibur's planned destruction of Prokaryon's planetary history, as it does for all purely social approaches to history, is a problem of biology. Biology places us within the cycle of living and dying, and as an "ageless" relativist, Nibur will repeatedly attempt to remove himself and others from that cycle. As he describes his own work with nanoplast, "I set men free—free from the limits of their flesh" (120). The already genetically perfected Elysian has dedicated his life to ensuring, through nanotechnology, that all other imperfections caused by basic human biology are overcome. But as Rabkin (1996) has suggested regarding the decadence of immortality, "When we put on incorruption, we are all changed: we are changed into ideals, into endless repetitions, into sterile vampires, childless angels, works of art, computer chips. We are changed into objects for the contemplation of others, but in the process we lose our very selves" (xvi). The aestheticization that immortality occasions, particularly in the manner in which Nibur accomplishes it, involves a loss of one's humanity. Rabkin does not discuss this loss of humanity in the often utopian manner of the cybernetic posthuman. He means it as a cautionary tale. To cite Gomel once again, this ahistorical impulse is a utopian one and "Utopias have their own heterotopias, which are institutionalized as necessary and concealed as shameful. They are known as concentration camps" (2014, 21). One cannot simply create an aesthetically incorrupt history, even an immortal one, without destroying the biology on which that history will be built.

Nibur seeks to step outside biology, and to a certain extent, his immortality allows him to do so. The decadence occasioned by his non-biology is also apparent in his desire to bring his pet dog with him into immortality: "Nibur, himself ageless, had had Banga lifeshaped before birth to be ageless like his master" (113). Age implicates the human body within biological cycles, but agelessness places Nibur outside their influence and limitations. What confounds Nibur and his plans to destroy Prokaryon's ecosystem, as well as the spatial metaphor that he uses to excuse this destruction, is the presence of living bodies, however micro they may be. Unfortunately, for Nibur, to destroy the building might mean destroying the occupant, and to the micros inhabiting the colonists, this is literally true. *The Children Star* likens ecocide (historical destruction of a planet) to homicide (biological destruction of a human body). Biological constraints break down purely constructivist positions, and the topic of age focuses on these limitations. For example, Marshall and Katz (2006) discuss how "today's youth-based and largely unachievable standards of successful aging and well-being are gaining prominence through the financial and cultural enterprises of agelessness that stretch the anxieties of middle age across the life course" (77). Nibur's is one such financial enterprise, and we can understand his motives for promoting "successful aging" as connected to his own anxieties over the biological. It is also important that in this context "successful aging" equates to "agelessness": What specifically foil his efforts, both figuratively and literally, are aged bodies.

Differently aged bodies draw us back into the biological cycle of living and dying. They force us to acknowledge the constraints that biology places upon even our intersectional social identities. Feminist gerontologists Calasanti and Slevin (2006b) note that "Proponents of 'agelessness' argue that being old is *all* a social construction … [that] we can avoid becoming old by simply not thinking and acting 'old'… [but] age categories have real consequences, and bodies—old bodies—matter" (4–5). As social theorists such as Butler (1993) have extensively documented, matter is certainly constrained by social and discursive forms of power. Matter constrains them in turn. Physical and biological constraints are not limited to the elderly. Although gerontologists position themselves against "youth-based" culture, children and the aging are treated similarly relative to the privilege of young adults in their twenties and thirties. What Nibur's agelessness amounts to, then, is a maintenance of the privilege afforded to this normative age group.

With their own unique biology foregrounding the process of aging, children provide a challenge to agelessness in *The Children Star*. Proponents of socially constructed "agelessness" have moved away from a kind of biological determinism that typically casts youth as problematic, becoming "increasingly disinterested in children's *becoming* (that is, what they might be) and far more interested in children's *being* (what they are)" and they stress "discursive constructions of social reality" (Best 2007, 10, 17).[6] In the gerontologist's critique of that age group, children are humans uniquely located within the biological process of aging. Childhood acquires meaning from the social values we ascribe to it, which also limits the extent to which we may treat children as "ageless." Rabkin (1996) lists among fiction's most common immortals "sterile vampires" and "childless angels." Not surprisingly, the Elysians are incapable of bearing children biologically, hence the shons in which their children are produced and reared apart from the rest of society. Whereas immortality or agelessness means stepping outside the biological process of aging, youth indicates a constant reentering of that cycle. For this reason, age is abjected from Elysian society and consciousness. But the biological will find ways to force even the ageless, the nearly immortal, to acknowledge their place within a material, lived history.

The Elysium Cycle astutely recognizes that a purely aesthetic history such as Nibur hopes to attain is linked to the exclusion, even the destruction, of the biological—both in the destruction of Prokaryon and in its "immortalization" in the ship Proteus, both of which involve childhood in different ways. The nature of Proteus is highly problematic for Nibur's immortalization. It is a servo—an intelligent machine. *Daughter of Elysium* dramatized the awakening of servos, their gaining of sentience and subsequent revolution. They continue to serve their human creators, but now they must be paid for their service. As one character remarks, "Of all the crimes of humanity ... this was the worst: that servos were created with no ultimate sense of purpose except to serve their destroyers" (1993, 329). Their "freedom" was hard-won, and that is why in *The Children Star* Proteus's condition as a "stable" servo is so controversial. Were this highly complex machine not intentionally restrained, it would very likely be sentient. The erasure of its nascent intelligence attracts a great deal of censure: "critics likened it to abortion, or even infanticide" (1998, 112). The text likens Proteus to a child on whose body Nibur repeatedly inscribes aesthetic histories. He will not be as successful with his planned destruction of Prokaryon.

Children such as 'jum, orphaned on their home worlds by poverty and disease, stand in stark contrast to the immortal Elysians. If Nibur gets his way, the rescued children on Prokaryon will be refugees, displaced as much as Elysian shonlings. But in several ways, children halt Nibur's planned destruction of the planet. It is 'jum, for example, who learns to communicate with the tiny native sentients of Prokaryon, who are the reason its destruction must be stopped. At one point, Mother Geode tells the children of the Spirit Colony a mythical story: "There was a world where only children dwelt. Where every creature grew with a saddle made for riding, and every tree formed little steps for climbing" (1998, 215). But this utopian children's world is also facing destruction: The star providing its warmth is growing larger and dying. Before the star dies, it tells the world not to fear, that it will never die, and "what the star said was true, for that world never died. It will live forever in the hearts of children who remember" (216). The immortal Elysians represent a process of historical spatial destruction; children represent an opposed biological organ of memory. "Children who remember" are bodies, physical structures, containing memories, firmly within the biological cycle of "living and dying."

That those children can be an unwanted reminder of mortality becomes especially apparent as they become analogs of the "hidden masters" themselves. Children may be small, but so are the micros. Continuing in the ecofeminist manner of the Elysium Cycle, the planet that the children occupy is just as biologically detailed as the human bodies in which the micros reside. But children are even more important to the story set on the infant world of Prokaryon. Prokaryon may not even be recognized by the Free Fold as an official settlement because "no colonist had yet borne a child live on Prokaryon… . There was no such thing as a native Prokaryan" (1998, 116). In an echo of Foucault's *History of Sexuality*, Prokaryon is not even allowed to enter the social order unless the Prokaryans are able to procreate (they, too, are at least temporarily "childless angels"). Much as feminist gerontologists suggest that old people are excluded from critical discussions because "they are not thought to have a productive future," the children of Prokaryon are excluded for not having proved their fertility (Allen and Walker 2006, 157). Children and the elderly have different sexualities, even while both are treated as "unproductive" and taboo.

However, *The Children Star* counters the abjection of age by aligning children with the most minor "sharers" imaginable, who not only turn out to have sentience but also turn the tables on their would-be colonizers by inhabiting their very bodies. Indeed, the destruction of Prokaryon is halted because even the immortal Elysians are subject to inhabitation by these micromen. The only solution is symbiosis and co-habitation on Prokaryon, because, as 'jum tells us, these creatures "need our help.... They're mostly children" (345). In fact, their status as human also has much to do with the familiarity of their care for their children. As the Elysian in charge of evaluating them decides, "their habits, their food, their children; to Verid all seemed uncannily familiar. These were people, no question about it. Immigrants to new worlds, only they found their new world was much more than a mindless landscape" (292). It is not the mere ability to procreate that grants them humanity, in this case, as it is for the human settlers of the planet. Rather, it is the care they take of their children and the respect they show to the life of their young. This "Children Star" is so named not only for the human orphans but also for the sentient creatures that inhabit their bodies.

This narrative of age therefore collides with and counteracts the revisionist narrative of spatial-historical destruction. Even the "micromen," like the children they inhabit in this novel, function as organs of memory, as that "intelligent pathogen who remembers its history and values its host" (238). In contrast to the Elysians, these intelligent microbes, mostly children, value physical bodies. As much as Prokaryon is to the humans, bodies are their planets. Valuing their hosts, however, is almost synonymous with remembering history. Because the microbes live their lives so much more quickly than their human hosts—one generation in a human hour—there is a very practical reason for their insistence on memory. The novel couples this memory, this history, with the notion of physical embodiment. "It is important to keep challenging the range of bodies that matter, so that these bodies will be taken into account when making ethical choices.... [I]t is necessary to retain a sense of embodied subjectivity, of real material consequences to our actions and choices" (Vint 2007, 187). The retaining of this "sense of embodied subjectivity" is in this case an act of memory, of historical preservation, and is necessary for ethical decision-making, which is precisely how Hughes-Warrington defines the act of historical revision. Regardless of their size, "children who remember" are bodies that matter in this text. In *Brain Plague*, which gives even greater voice to the micros, the biological underpinnings of social constructions acquire even more potency.

A Plague on Both Your Houses

Brain Plague (2000) gives a clearer voice to this race of intelligent microbes, in particular to the micro-culture known as Eleutheria. It is important to distinguish micro-cultures, because in this novel numerous human characters have become carriers and each human's "people" develop their own cultural identities. Chrys carries the people of Eleutheria, a libertine society of architects or, because they build with the intelligent nanoplast, "dynatects." They have constructed some of the largest and most impressive buildings of the era (e.g., the "Comb," a massive sentient building), although their existence is threatened both by human suspicion that they may be controlling people's minds and by the dark microbes of Endless Light who enslave their human hosts. There is a hint, however, that all cultures, good and bad alike, have a kind of affinity with building, as "plague micros built the Slave World, just as ours built the Comb" (2000, 22). Yet the enslaving micros destroy their human hosts and, thus, themselves. Symbiosis is especially important now, because humans are now all potentially "children stars" as they host the microbial races and must learn to negotiate with them. The microbes, in turn, construct vast cities at the base of the brain stems of their hosts, building with proteins their houses, nightclubs, and even jails. Like the giant buildings of nanoplast constructed in the outside world, these microbial metropolises represent a careful amalgam of biology, history, and physical place.

Both *The Children Star* and *Brain Plague* offer stories of attempted colonizations and the negotiations needed to cultivate cooperation between different peoples. In both, this process is threatened by a relativist destruction of planetary history that equates to destruction of the human body. Thus, when Chrys takes on the Eleutherian society of micros, she is already aware of the dangers. Most people think of the micros as the plague that has given rise to entire underworld populations of "vampires," rabid people controlled by bad micros, infected addicts whose bite transfers and spreads their micro-society. Chrys also faces dangers from humans who know only that she carries the Eleutherians, who work as dynatects on maintenance of the "Comb." In fact, their prior host, also famous for this work, was killed for being a carrier. Her new artworks bring her renown, as she begins to draw the images of the micros who communicate with her.

Most of the plot centers on efforts among the Carriers to regulate their human–micro relationships. Chrys and her people become inspectors, checking other humans and their people for potential defection to the slave micro-society known as Endless Light. She and the other carriers will chase these slave micros (who infect bodies in order to starve them, discard them, and move on), root them out of even the perfect Elysians' bodies, and eventually either convert or destroy them. With the looming threat of these slave micros' readiness to discard human bodies like so many dead planets, so intimately connected to the preservation of history, it is the Eleutherians' internal building that is of special importance in the context of history. Like the Sharers and their rafts earlier in the Elysium Cycle, the Eleutherians also go to great lengths to preserve their collective history within the structures of their habitat. Their accelerated lives force them to take such measures: "Memory, Unseen, is the most sacred light of Eleutheria. Memory marks us worthy of the Blind God's promise; worthy to dwell with a new god, for whom our generation lasts but a day. Tell the children: Always remember" (18). Because their cycle of living and dying is so brief, and because it stands in such direct contrast to that of their human hosts, they must develop a strong collective memory. Thus, the few elders of the community also recognize the importance of "children who remember" and are careful to tell their children their history and to codify it within the walls of the cities they build. This kind of history, built out of a necessity driven by the very obvious mortality of the microbes—indeed several generations die in the few weeks of the novel's plot—is precisely what Nibur sought to destroy in *The Children Star*. And for all their talk of immortality, Elysians are careful to make a distinction: "We never say 'immortal,' only 'ageless.' Most Elysians who escape accident will make it to a thousand, but a small percentage won't. Nothing's perfect" (1993, 34). Nibur, too, is all too aware of his vulnerability—even his lifeshaped dog dies of the "plague." Just as Nibur's destruction of the biological history of a planet is driven by his having left the cycle of living and dying, the micros' biological preservation of history celebrates an acceptance of their mortality or, rather, the natural process of aging.

This preservation is a continual process. Against the determinism of empirical truths, constructivists stress the notion of process at the base of becoming. The Elysium Cycle encodes this process within the biological mechanisms of historical preservation: "For in truth, Eleutheria is no genetic race, nor a physical place, but a way of being, a path of endless life. All those who seek to build in truth and memory shall find our way" (2000, 294). Cultural identity itself is built upon the careful transmission of history from one biological organism to another. A relative historical perspective privileges the individual subject in the present moment, but it also tends to forget physical bodies. The past inevitably involves the deaths of the many before the individual. To forget one's own mortality is to jettison that past.

The micros anticipate and foreground the importance of age as it enters into larger critical discussions. In particular, scholars of age studies recognize the need to balance (not to negate) the social-constructivist position with the notion of physical, biological embodiment. Thus, Laz (2003) has posited that we *do* age, much as we *do* or *accomplish* gender. However, one must also recognize constraints on this performativity, because "we make our own bodies but do not make them just as we please; we do not make them under circumstances chosen by ourselves but under circumstances found, given, and transmitted from the past" (507). Moreover, biology and physical embodiment are some of the many constraints that we all encounter. Therefore, "how one 'does' age has implications for corporeal experience," just as "embodiment has implications for how one can accomplish age" (508). In the case of the micros, their very brief lifespans led primarily as children are the biological constraints that confront their entire society. They are circumstances "transmitted from the past," and their methods of confronting them are not to negate them but to embrace history—to *do age* naturally.

To do age naturally should not be taken as yielding to biological determinism. There is no single way to do one's age. Similarly, there are many different cultures of intelligent microbes, each with their own interests and strengths, and they constantly change and adapt. There are even schisms within the Eleutherian society: Some children decide they do not want to live as the other Eleutherians. There is no one "proper" way to age, but it is vital to consider the biological implications of age, even alongside its social construction. There are strengths to ages, both old and young, but also limitations and constraints, many of them biological. Not to acknowledge these limitations, that is, to place one's self outside the cycle of living and dying as Nibur has done, may constitute a destructive ahistoricism. To service the denial of his own mortality, Nibur would entirely destroy the planet Prokaryon, an unethical act of forgetting tantamount to the destruction of human biology.[7] To forget the biological component of humanity signals what Agamben (2004) calls the "decisive political conflict, which governs every other conflict… that between the animality and the humanity of man" (80).

Moreover, the material stakes of this forgetting counsel us to read history as equally embodied. The micros demonstrate the relationship between written history, body, and place, but the culture of Eleutheria in particular values history and the body. In contrast, the plague micros—those that enslave their human hosts by accessing their pleasure centers directly—mirror the decadent agelessness of Nibur. While Nibur is thwarted in his planned ecocide, the plague micros demonstrate the material, embodied dangers of subjugating history to private desires. In order to save her lover, Daeren, Chrys visits the slave world and sees firsthand their debasement of the human body. Unlike Eleutheria, Endless Light sees the human body as a mere receptacle, a raw material to be used and discarded. Thus, Chrys sees many "worker slaves, their eyes all flickering white as they passed. Some pushed cots or wheelchairs containing human bodies, inert, with unkempt beards or bare breasts, eyes horribly staring" (2000, 247). And when she finally finds Daeren, now literally occupied by the leader of Endless Light, she bargains to get him back at all cost. For her it is "better to take his empty shell than to have to see him in her dreams, as he would eventually be, his body exposed to unspeakable decay" (248). On the slave world, the plague micros have no need to care for the human bodies of their hosts, because they have no desire to stay within the boundaries of a material history. In fact, it is their explicit desire to step outside both biology and history.

Chrys succeeds in rescuing Daeren by offering the leader of Endless Light a kind of immortality (nor is it accidental that "slave workers" are also often referred to as "vampires"). She shows him a portrait she made of one of her micros, Rose, and the leader demands that she paint his own portrait to be put on display for everyone "*to see my enlightened form raised before all*" (249; emphasis in original). Just as Nibur decadently "immortalizes" planetary ecosystems on the shell of Proteus's restrained sentience, the plague micros seek a kind of immortality for which they are clearly willing to destroy planet and body. Biological processes of aging and material written history encoded in the bodies of their human hosts, the plague micros are willing to destroy the human body to achieve immortality, to forget about the biological cycle of living and dying. The slave world is like an entropic description of an elderly community with no "productive" future. Mere bodies are used up and ready to discard. To the living history of Eleutheria, the plague micros oppose a constructed agelessness that highlights the clear cost of this position as the degradation of the human body.

Ultimately, the Elysium Cycle highlights the importance of remembering this abject body. Remembering to include the biological along with the social also means to include the human forms of age within the cycle of living and dying. Our wish to abject age in order to "encourage preferable dreams" (Southgate 2009, 200) as Southgate suggests, whether of immortality or something else, could destroy our definition of humanity. There is a powerful message here for the critic of both history and age. To paraphrase Hughes-Warrington, even though revisionist histories give the appearance of new boundaries not to be crossed, the fluidity with which revisionist histories engage them, the free and open dreaming of reconstructing and revising history, may also offer the critic a means to "interrupt" their revision. That is, if history may be constructed at will, the critic may highlight and critique that very process in order to disrupt the unethical revision of history. The tiny micromen are children who remember, and they disrupt destructive history-makers by violating the boundaries of their physical bodies. The bodies they inhabit also remember—they have a history as a physical place. Chrys not only saves Daeren, but the Eleutherians save Daeren's people, the Blue Angels, and bring them back to "build in truth and memory." The careful critic ought to do the same by insisting on the physical, the biological, alongside the social.

My reading of these novels suggests that a kind of cautious self-reflexivity is important in the adoption of purely social models in critical work on science, history, and even science fiction. Social criticism corresponding to the cultural-linguistic turn now prevalent in historical philosophy is now also a common approach in SF criticism dealing with time and history. Hills (2009), for example, declares that "sf's interest in possible worlds is one which resonates with life in late modernity … where risks are variously calculated, projected, and themselves managed as counterfactuals" (440). And there are certainly good reasons for this type of counterfactual "management," as SF has often found itself challenging established histories that serve to marginalize particular groups. The importance of feminist science fiction is one such example, and Yaszek (2009) describes how SF authors "use alternate histories to explore how science and society might evolve outside Western paradigms" (197). The importance of "history free and experimental" is that it has allowed writers and critics to bring much-needed attention to neglected issues of race and gender and to allow for the rediscovery of historical contributions of groups not limited to the "West" (i.e., white, male, of European descent). This importance cannot be overstated. However, without losing sight of these vital aspects of the counterfactual play with history, it is also possible to ask what might be lost in our enthusiasm for social and historical constructivism.

In the scholarship of aging, the tendency toward constructivism is youth-centric and does not account for the unique needs that aging bodies place upon us. The popularity of this approach, therefore, implicates our own aversion to aging and our "preferable dreams" to step outside of a pattern that ends with dying. In this sense, then, it is not surprising that SF criticism may favor the constructivist approach. Vint (2007) cautions that "to avoid the dangers of abstraction, it is important to give representation to the *full range* of human embodiment," including the intersections of race, class, gender, age, social class, etc. (184; emphasis added). As scholars in age studies have argued, however, criticism in any field (and SF, with the exception of a recent special issue of *Femspec* dedicated to "Great Age," does this as well) has greatly neglected the topic of age, a constructed relegation to the social margins. This is not unusual—one may both acknowledge that age studies is a fairly new discipline and also note that SF joins most other critical approaches in neglecting the topic of age for treatments of race and gender. I hope to have identified a potentially rich and largely untapped area for deeper consideration.

Slonczewski's work does not negate the importance of the social construction of categories such as race and gender, so well documented in other critics' work, but still allows for a critical intervention, sensitive to the embodied, physical limits and consequences of relativist and constructivist approaches in our critical considerations. Bodies matter, and so do the physical embodiments of history, place, and person. Favoring approaches that emphasize the contingency of history and biology deemphasizes the limits that history or biology place upon us. This essay has argued that a careful recognition of age revitalizes notions of physical, grounded history and biology. The negation of a particular history, the revision of that history, or the act of forgetting may allow us to encourage "preferable dreams" or to create a utopian space in which to maintain our psychological stability in the face of troubling realities, such as growing old and dying. Certainly, it serves those such as Nibur and the leader of Endless Light, who dissolve history into decadent personal aesthetics. In contrast, the biological and historical bodies of the Sharers, the carriers, and the micros negotiate lifeways that avert such destruction.

Notes

1. Age studies scholars frequently highlight cosmetics and pharmaceuticals (i.e., Viagra) as signs of a widespread social pressure to conform to standards of youth.
2. Elsewhere, I have questioned why Foucault, in relating the story of a young girl and a "rather simple" village man playing a game of "curdled milk," descries the proliferation of discourses of sexuality surrounding the man but does not deign to consider the absence of discourse surrounding the girl's experience. See Thiess (2015).
3. One must also note that "myth" plays an important role in feminist approaches to SF, allowing the inclusion of "herstory" within history and stressing indeterminacy over deterministic scientific approaches. Roberts (1990) compares the Sharers' science in *A Door into Ocean* to magic and witchcraft, stressing that "The Sharers' science is linked to language, and this conjunction stresses a non-hierarchical view of the universe" (140). The positive way that such indeterminacy allows for reconsideration of gender norms is important.
4. There is a wealth of scholarship on alternate histories (e.g., Hellekson 2001). My concerns differ from this body of scholarship in that I am interested in focusing on questions of relativist philosophies of history rather than on the links between narrative and history.

5. There are hints of this in earlier novels in the Elysium Cycle. *A Door into Ocean* tells the story of the colonization of this moon and the non-violent resistance of the Sharers who, because they are accustomed to their homes on floating rafts being regularly destroyed, seem to stand apart from the cycle of historical place destruction. In fact, they become protectors against ecocide in the future novels, protesting and "witnessing" the potential destruction of Prokaryon; they also serve as organs of memory for those planets they are unable to save. In *Daughter of Elysium*, the "Sharer delegates never let Raincloud [the protagonist from planet Bronze Sky] forget that she inhabited a murdered planet, its ecosphere erased for the sake of human habitation" (42). The novel suggests a potential solution to this violent cycle in an insistence on remembering, or not forgetting, the violent displacements of the past.
6. One can see the logical ends of this movement even in SF criticism. For example, Lehtonen (2013), perhaps the clearest treatment of age in either SF or fantasy to date, primarily examines the subversion of age. Like gender or race, it is merely another boundary to be transgressed.
7. One might arguably challenge here whether Nibur's attempted act is wrong in and of itself, or whether it is only wrong because the planet contains sentient species, and, likewise, whether it would be wrong if there were no such species. What Slonczewski's work indicates, however, is that the very alien nature of such sentience makes it difficult to discover in the first place. Characters often wonder what sorts of life might have been destroyed in hastily boiling off other planets in the Free Fold.

References

Agamben, Giorgio. 2004. *The Open: Man and Animal.* Trans. Kevin Attell. Stanford: Stanford University Press.

Allen, Katherine R., and Alexis J. Walker. 2006. Aging and Gender in Families: A Very Grand Opening. In *Age Matters! Realigning Feminist Thinking*, ed. Toni M. Calasanti and Kathleen F. Slevin, 155–74. New York: Routledge.

Attebery, Brian. 2002. *Decoding Gender in Science Fiction*. New York: Routledge.

Best, Amy L. 2007. Introduction. In *Representing Youth: Methodological Issues in Critical Youth Studies*, ed. Amy L. Best, 1–36. New York: New York University Press.

Bollinger, Laurel. 2010. Symbiogenesis, Selfhood, and Science Fiction. *Science Fiction Studies* 37 (1): 34–53.

Butler, Judith. 1993. *Bodies That Matter: On the Discursive Limits of "Sex"*. New York: Routledge.

Butter, Michael. 2009. *The Epitome of Evil: Hitler in American Fiction, 1939–2002.* New York: Palgrave Macmillan.

Calasanti, Toni M., and Kathleen F. Slevin, eds. 2006a. *Age Matters!: Realigning Feminist Thinking*. New York: Routledge.

———. 2006b. Introduction: Age Matters. In *Age Matters! Realigning Feminist Thinking*, 1–17. New York: Routledge.

Csicsery-Ronay Jr., Istvan. 2008. *The Seven Beauties of Science Fiction*. Middletown: Wesleyan University Press.

Gomel, Elana. 2014. *Narrative Space and Time: Representing Impossible Topologies in Literature*. New York: Routledge.

Hellekson, Karen. 2001. *The Alternate History: Refiguring Historical Time*. Kent: Kent State University Press.

Hills, Matt. 2009. Time, Possible Worlds, and Counterfactuals. In *The Routledge Companion to Science Fiction*, ed. Mark Bould, Andrew M. Butler, Adam Roberts, and Sherryl Vint, 433–41. New York: Routledge.

Hughes-Warrington, Marnie. 2013. *Revisionist Histories*. New York: Routledge.

Foucault, Michel. 1990. *The History of Sexuality, Volume One: An Introduction*. Trans. Robert Hurley. New York: Vintage.

Laz, Cheryl. 2003. Age Embodied. *Journal of Aging Studies* 17 (4): 503–19.

Le Guin, Ursula K. 1973. The Ones Who Walk Away from Omelas. In *The Wind's Twelve Quarters*, ed. Ursula K. Le Guin, 275–84. New York: Harper and Row.

Lehtonen, Sarah. 2013. *Girls Transforming: Invisibility and Age-Shifting in Children's Fantasy Fiction Since the 1970s*. Jefferson, NC: McFarland.

Marshall, Barbara L., and Stephen Katz. 2006. From Androgyny to Androgens: Resexing the Aging Body. In *Age Matters! Realigning Feminist Thinking*, ed. Toni M. Calasanti and Kathleen F. Slevin, 75–97. New York: Routledge.

Rabkin, Eric S. 1996. Introduction: Immortality: The Self-Defeating Fantasy. In *Immortal Engines: Life Extension and Immortality in Science Fiction and Fantasy*, ed. Gary Westfahl, George Slussar, and Eric S. Rabkin, ix–xvii. Athens: University of Georgia Press.

Roberts, Robin. 1990. Postmodernism and Feminist Science Fiction. *Science Fiction Studies* 17 (2): 136–52.

Scarry, Elaine. 1985. *The Body in Pain: The Making and Unmaking of the World*. Oxford: Oxford University Press.

Slonczewski, Joan. 1986. *A Door into Ocean*. New York: Orb/Tor, 2000.

———. 1993. *Daughter of Elysium*. Rockville, MD: Phoenix Pick, 2010.

———. 1998. *The Children Star*. New York: Tor.

———. 2000. *Brain Plague*. Rockville, MD: Phoenix Pick, 2010.

Southgate, Beverley. 2009. *History Meets Fiction*. London: Longman Pearson.

Thiess, Derek J. 2014. *Relativism, Alternate History, and the Forgetful Reader: Reading Science Fiction and Historiography*. Lanham, MD: Lexington.

———. 2015. *Embodying Gender and Age in Speculative Fiction: A Biopsychosocial Approach*. New York: Routledge.

Vint, Sherryl. 2010. Animal Studies in the Era of Biopower. *Science Fiction Studies* 37 (3, November): 444–55.
———. 2007. *Bodies of Tomorrow: Technology, Subjectivity, Science Fiction.* Toronto: University of Toronto Press, 2010.
Wolmark, Jenny. 2005. Time and Identity in Feminist Science Fiction. In *A Companion to Science Fiction*, ed. David Seed, 156–70. Malden, MA: Blackwell.
Yaszek, Lisa. 2009. Cultural History. In *The Routledge Companion to Science Fiction*, ed. Mark Bould, Andrew M. Butler, Adam Roberts, and Sherryl Vint, 194–203. New York: Routledge.

CHAPTER 6

Microbial Life and Posthuman Ethics from *The Children Star* to *The Highest Frontier*

Sherryl Vint

> In protected homes across the empire, humans have curled up in their armchairs with their pets and their species-simulated snacks to watch the destruction of the rest of the world on TV. It is hard to know whether any humans will survive such domestic dreams. Fungi are not taking a position. Even the hardy lichens are dying from air pollution and acid rain. When they take up radioactivity from nuclear accidents, they feed it to the reindeer, who in turn feed it to human herders. We can ignore them, or we can consider what they are telling us about the human condition.—Anna Tsing, "Unruly Edges"

Rethinking our species beyond the limiting frameworks of the human and into the expanse of the posthuman has become a central focus of scholarship in the humanities, much of it attentive to our entanglement with the lives of other species. Work by scholars such as Haraway (1991, 2007) and Wolfe (2009) points to the centrality of the human/animal divide in

S. Vint (✉)
Media and Cultural Studies Department, UC Riverside, Riverside, CA, USA

© The Author(s) 2020

B. Clarke (ed.), *Posthuman Biopolitics*,
Palgrave Studies in Science and Popular Culture,
https://doi.org/10.1007/978-3-030-36486-1_6

requiring us to rethink a prior philosophical and political model of the human that has largely been premised on disavowing our myriad connections with animal life. At the same time, post-genomic developments in biotechnology repeatedly demonstrate that elements of the human can be recombined with other species, that human bodies can be assisted and augmented with materials taken from non-human animals, and even that entities such as research cell lines, derived from human bodies, can have lives that far exceed those of their donor bodies. What we mean by "human" is no longer self-evident, and rethinking this human condition has tremendous potential for transforming our understandings of politics and ethics.

Yet, as Braidotti (2013) points out, whether our willingness to reconceptualize the human outside of pedantic frameworks about the "natural" will result "in playful experimentations with the boundaries of perfectibility of the body, in moral panic about the disruption of centuries-old beliefs about human 'nature' or in exploitative and profit-minded pursuit of genetic and neural capital, remains … to be seen" (2). Much that has been written about the posthuman has focused on two main sites of enquiry. One is related to the human body, augmented by machine prostheses or by genetic engineering, perhaps bypassed entirely so that minds could be uploaded to digital existence. The other is focused on humanity as one animal among many, part of complex ecosystems and no longer privileged in the possession of some capacity (reason, language, a sense of time, sentience, sapience) that would justify its place above all other species. I will focus on the latter. As the epigraph from Anna Tsing denotes, beginning to think of humanity and the effect of our choices on all living species requires not only that we begin to care for and about other species through such entanglements, but also that we recognize the degree to which the consequences of such choices remake the human as posthuman, perhaps in unanticipated ways. Indeed, becoming posthuman in this way may well be necessary for what Tsing (2012) calls "collaborative survival," an orientation she defines as the counterpoint to the sort of zero-sum game ethics of survivalism now ubiquitous in popular culture. For Tsing, "staying alive—for every species—requires livable collaborations. Collaboration means working across difference, which leads to contamination. Without collaborations, we all die" (28).

In her foundational "Cyborg Manifesto," Haraway (1991) asked, "Why should our bodies end at the skin, or include at best other beings encapsulated by skin?" (178). Inspired by Slonczewski's fiction and its complex representation of our myriad interactions with microbial life, I want to think here about posthumanism through our skin, a posthuman emerging from the multiplicity of other lifeforms that inhabit human bodies, living on the surface and within, and in many ways constituting the human body itself. Thinking about the posthuman community of human/microbe co-existence as theorized in this work offers a compelling model of Tsing's ethics of collaborative survival. Moreover, this fictional model allegorizes and makes visible a material reality in which human lives and bodies are in continual interaction with a wide range of microbes, good and bad, a vision informed by Slonczewski's research on microbial life.

Microbial Political Life

Theorists have recently re-evaluated the way that we think about microbes, concurrent with shifting paradigms in medical research conceptualizing the human as a supra-individual, that is, as an organism made up of the human body and the microbiome that lives within and on it.[1] Reviewing the work of Lynn Margulis, Hird (2009) argues that life itself is fundamentally social and that this foundation of exchange and interaction begins at the microbial level. Theorizing based on the difference between prokaryotic cells such as bacteria (whose genes are not enclosed in a membrane-bound nucleus) and eukaryotic cells (like our own, which have this structure), Hird points out that eukaryotic cells almost without exception contain subcellular structures that were once independent prokaryotic cells. Thus, it is not merely that we live surrounded by microbes or that some microbes enter other animal bodies and enact change within them, but further that the cells of these bodies, which we think of as "human" in our case, themselves contain non-human prokaryotic life that has become absorbed into them, parts that are necessary for their proper function. For Hird, life is thus fundamentally social and symbiotic, necessary a group of distinct entities who must cooperate across their differences for the survival of all. Similarly, Paxson (2008) encourages us to remember that there are many more "good" than "bad" microbes, and, for all the ones we must see as potential health risks that will cause disease in human bodies, others such as bacteria, yeast, and mold regularly transform milk proteins into nourishing foods such as yogurt or, her object of study, cheese.[2]

Landecker (2016) asks us to think etymologically about the word "anti-biotic" as a version of "anti-life." Responding to concerns about the approach of an era when microbial mutation means that none of our current antibiotic drugs retain their effectiveness, Landecker questions the politics of the era of wide-ranging use of broad-spectrum antibiotics. Her work reveals that the use of antibiotics emerges from a different conceptualization of the human/microbe relationship than the one championed by Hird, competition rather than cooperation. Antibiotics "kill by selective toxicity, disrupting microbial structures or processes that do not exist in human cells. Their production is driven by theories of antibiosis: a human leveraging of substances microbes create in mutually antagonistic battles for space and resources" (2). Medicine has embraced this agonistic model, and one might even suggest that an instrumental way of framing the relationship between humans and microbes has contributed to the rapid arrival of an era of antibiotic resistance.

Landecker's work traces how substances with antibiotic properties are found in narrow and specific contexts, where a history of microbial interaction has produced them, but are then taken into sites of industrial manufacture and distributed over vast territories they would not have reached without such human intervention. Moreover, Shapiro (2015) describes the bacterium *E. coli* as able to "engineer its own genome" (176); drawing on Barbara McClintock's work, he notes several ways that cells can repair damaged chromosomes or otherwise respond to environmental factors, which includes horizontal DNA transfer among bacteria of sequences that convey antibiotic resistance. Noting that they can also acquire DNA from their environment, including from viruses, he dubs bacteria altogether "the smartest cell biologists on the planet" who exhibit a control of events "in cells of higher organisms" that "mere human scientists can only dream of imitating" (189).

By tracking the horizontal gene transfer through which microbial life exchanges genetic information and capacities with other microbes, we can see what Landecker calls the biology of history: "how human historical events and processes have materialized as biological events and processes and ecologies" (21). In the case of microbes, this biology of history includes "the historical record of human antibiotic use inscribed across the biology of bacteria" and "historical environmental events that drive the evolution of pathogens and commensal bacteria alike, in bodies and far beyond them" (21). Landecker notes that the plasmid—"a small extra-chromosomal circle of DNA that moves between bacteria during conjugation" (29)—enables bacteria to transfer resistance to one another. This circular image is central to how the microbes are envisioned in Slonczewski's work, beginning with her "first contact" novel, *The Children Star* (1998). The contact in question is with a species of alien life, but one that is microbial and communal, living through generations at a vastly different timescale than the time of human decision-making, which greatly complicates efforts to establish communication and, eventually, cooperation between them. Landecker (2016) observes that "bacteria today have different plasmids and traits and interrelations and capacities and distributions and temporalities than bacteria before modern antibiotics" (21), and Slonczewski's science fiction makes this evolution of microbial communities—and their interactions with humans, both beneficial and dangerous—visible on the scale of human perception and temporality.

The Children Star

On the planet Prokaryon, all lifeforms, although "macro-sized"—that is, visible to human eyes—are prokaryotic, with cells whose genetic information is not contained within a membrane wall. The novel imagines a rich ecosystem of such prokaryotic lifeforms: "the higher structure of all the multicellular organisms was toroid, from the photosynthetic 'phycoids' that grew tall as trees, to the tire-shaped 'zoöids' that rolled over the fields they grazed—or preyed upon those that did" (1998, 8). This description of prokaryotic lifeforms through language that is familiar for conceptualizing a wide range of eukaryotic lifeforms, from plants to herbivores to predators—and we could go on to humans—can prompt us to recognize that in the diversity of the bacterial world beyond the novel, in microbial life altogether, most microbes are not "germs."

The humans who come to the planet are part of the Free Fold, an interstellar collective whose politics form the background to the story. Rival factions in the Free Fold struggle over the future of Prokaryon, a planet whose unique, spherical, and prokaryotic biology is poisonous to humans. Some wish to terraform the planet as a space for human colonization, relieving the suffering caused by overcrowding on many other planets. The Free Fold requires that the claims of sentient races to their home planet be respected, should such a race or species be found, and so a research team on Prokaryon searches for evidence of what they call the "secret masters" of the planet. They see strong evidence of cultivation and other shaping of the planet's ecology toward certain productive ends, yet they literally cannot see any species that play such a role—because this race is microbial, living on and within larger lifeforms. Yet another perspective, also on the side of terraforming, is found in the capitalist and colonial Elysians, represented by billionaire Nibur Lethe*shon*, who tries to buy fifty-one percent of the planet's assets so that he can control the harvest of its resources, rare minerals that must be mined in ways destructive to existing ecology. The creatures of Prokaryon have ring-shaped cells and a biology based on arsenic, making the planet poisonous to humans. The research team has been lifeshaped to enable them to survive on the planet, but this expensive and time-consuming process has kept their numbers small.

The novel begins from the perspective of an impoverished and orphaned child on the planet L'li, a space beset by inequity, overcrowding, and poverty, complicated by a prion disease that orphans this girl, 'jum. The research team on Prokaryon is staffed by members of the Sacred Order of the Spirit, an organization with values like Quakers: As they describe it, "Members of our order worship the Spirit of the land, wherever we dwell" (65). One member, Rod, rescues 'jum from L'li and takes her to Prokaryon, where the Order has established a colony with a disproportionate number of such rescued children because their bodies are better able to accept the massive lifeshaping changes required to live in the arsenic-based environment. Committed to preventing the terraforming of Prokaryon, the Order searches for the intelligent life whose presence would prompt a legal ban on mining and terraforming. One of Rod's colleagues believes that the entire planet is a single organism that communicates by exchanging microzoöids. As she points out, this theory is not as different from earth-based biology as it might first appear. "You could say we 'communicate' through our bacteria" (32), she tells Rod and hopes that when she begins looking at microzoöids she will at last discover the hidden masters of Prokaryon. Prokaryon hosts complex interactions as microbes are exchanged between human bodies, the environments of those bodies, and the environment, forming a kind of communication among humans, microbes, and other environmental factors.[3]

This communication is challenging, however, because the kinds of chemical signals passed across lifeforms can communicate on the level of biological response, but this channel of information does not allow for the complex phrasing that we think of as a language. The researchers discover that distinct colonies of microbes display responses of different colors when put into contact, and they are also able to produce a signal of flashing light, which 'jum, who proves to be a math prodigy, notices follow a sequence related to prime numbers. The microbes infect Rod, but at first see him only as another, lesser species, like the other lifeforms they inhabit on the planet, and they try to direct him to follow their will, through manipulations of the pleasure and pain centers of his brain. Eventually, however, as other authors in this volume have also noted, they use a system of flashing light to communicate more complexly with him.

In this charged scene of interspecies contact, Rod teaches them the human alphabet through a set of interactions that use physical materials to model the words in response to patterns in the flashing light, yet even with more complex exchanges, initially they continue to direct him rather than consult him. Rod and the microbes must learn and evolve together to form a more symbiotic supra-individual. The title for the book and the colloquial name for the planet, the Children Star, refers doubly to the human children lifeshaped by the Order to live on the planet and to the cycle of microbial life: Their generations change much more rapidly than do human ones, and their reproduction through three-stranded DNA does not leave the parent entities intact after the daughters are produced. Thus, much like the Order of the Spirit colony, microbial communities are mainly children with a few elders who remain to guide them.

As the Order works to identify and then communicate with this intelligent species on Prokaryon, struggles continue in the wider Free Fold regarding the planet's fate. Nibur plans to raze the planet to extract its minerals, terraforming what remains for colonialization, and he insists that the virtual reproduction of its landscape that he will create on his star-ship is equivalent to the material one, thus making this destruction no loss at all. His limitations speak to the novel's ecological themes. The Order tries to inform the rest of the Free Fold about the microbes' existence, but most are resistant to thinking that microbial life could be either intelligent or communicative, and respond with brute violence. When it is discovered that the microbes can infect Terran species, for example—the first step toward communication—the Free Fold's response is to "incinerate all nonhuman potential carriers" (157), namely the animals kept by the colony.[4] The microbes, too, need to be taught to respect the autonomy of their human hosts, to truly communicate rather than coerce, whether this be via pain or via addicting the host to pleasurable chemicals. The microbes inhabiting 'jum develop the most complex cross-species interactions and are able to tell her stories of their culture and evolution, the struggle of many generations to migrate to the harsh new landscape of her foreign body and learn to live in this new world. These adventurous microbes enjoy the challenge of inhabiting a world with its own agency, not the "mindless" world of their animal ancestors. They love their new world—that is, 'jum—proclaiming, "we discovered an amazing thing: the alien world had intelligent feelings. You could, in fact, understand us, responding to our most intimate desires. And most astounding of all, you came from the stars, like the very gods. It is a wondrous thing, to inhabit a god" (176).

Prokaryon is saved from terraforming due to the communication with the microbes, which proves that there is sentient life on the planet. The microbes can adjust human bodies to live on their world much more quickly, and less expensively, than the previous technologies of lifeshaping, and so the world is not entirely safe from change, but the Free Fold votes against the genocidal solution of terraforming. Microbe communities continue to develop and take on traits associated with their human hosts, a voluntary arrangement, truly creating a new kind of symbiotic identity for both, and many microbes are excited to travel to the stars on these new and more mobile worlds. At the same time, however, just as there is diversity of opinion among the humans, so too are their dissenting factions among the microbes. Not all are compelled by the idea of inhabiting an intelligent world. Some contend it is "TOO DANGEROUS. NOT WORTH THE RISK," which dismays Rod even as he reflects that this attitude is similar to that of fearful humans who incinerated the llamas.[5] On their own, the humans cannot defend themselves against these harmful microbe factions, but Rod's resident microbes promise him that such treatment of their intelligent worlds would be detrimental for the new, symbiotic microbe communities. They promise, "WE WILL PROTECT YOU, BUT OUR BROTHERS WILL KILL US, TOO. FAR MORE OF US WILL DIE THAN YOU" (188). The human–microbe supra-individuals that emerge at the novel's end are an illustration of the way that helpful and harmful microbes interact within the human body, encouraging us not to adopt the anti-life attitude of an antibiotic culture that indiscriminately seeks to wipe out microbes. They are also an image of the kind of collaborative survival across species evoked by Tsing: Through their communication, the physical world of Prokaryon is saved from terraforming and the microbial worlds of many colonies—that is, the human hosts they inhabit—are saved from the damage the arsenic environment would otherwise enact on them.

Microbiopolitics

The human discovery of and response to microbes can be understood as one element in the mode of governance that Foucault (1978) has characterized as the biopolitical, that is, the governance of the very biological life of the (human) species, including the emergence of regulations around public health, which include regimes to manage microbial infection. Latour (1993) suggests that it was hygienists' control of microbes that created the paradigm for rationally controlled social relations that could be directed toward some planned end, the epitome of biopolitical management. Microbial life interrupts and diverts this human biopolitical governance, and the elimination of microbes through sterilization and similar practices enables humans to assume complete control over social and physical worlds:

> If we wish to realize the dream of the sociologists, the economists, the psychologists—that is, to obtain relations that nothing will divert—we must divert the microbes so that they will no longer intervene in relations everywhere. They and their ways must be interrupted. After the Pasteurians have invaded surgery, only then will the surgeon be alone with his patient. After we have found a method of pasteurizing beer, then the brewer will be able to have nothing but economic relations with his customers. After we have sterilized milk by spreading throughout all farms methods of pasteurization, then we will be able to feed our infant in a pure loving relationship. (Latour 1993, 39)

Latour's point is this passage is not merely the antagonistic frame for our interactions with microbes, but more centrally that we must recognize them as actors in the coproduction of reality, even if their agency is expressed only through the negative framing of "diversion."

Building on Latour, but also articulating a more complex vision of microbial life that acknowledges its variety, Paxson (2014) outlines a theory of what she called "microbiopolitics," beginning from the premise that "dissent over how to live with microorganisms reflects disagreement about how humans ought to live with one another" (16). Reading the practice of cheese cultivation in Foucauldian terms, Paxson suggests that the effort put into caring for the cheese—seeding milk with a microbial culture that will produce the desired taste and consistency, protecting the curds from contamination by other microbes, storing the aging cheese in a way that allows it to develop a protective rind—can also be literally understood as a care of the self: "what protects the cheese can protect us" (32). Through cheese, humans and microbes create a cooperative culture that produces food for both humans and the "good" microbes, a place to live for the microbes, and the probiotic benefits of cheese consumption for human health. This image of a cooperative community between humans and microbes is the opposite of Pasteurian practices, which foster life by killing microbes. Using language similar to Tsing's notion of collaborative survival, Paxson argues that artisanal cheese production is a "collaborative human and microbial cultural practic[e]" (17).

Paxson does not explore in detail this connection between Pasteurian hygiene and Foucault's formulation of a new operation of biopower in disciplinary societies. In the context of Slonczewski's work, however, and its rethinking of human and microbial communities as a kind of posthuman supra-individual, it is worth revisiting how it is that a politics of life becomes simultaneously a politics of death. The sovereign's power over life and death was originally expressed, Foucault (1978) theorized, in the sovereign's right to take the life of anyone who threatens him, but as sovereignty migrated from the power of an individual ruler into an expression of the will of the social body, this power of death became "the reverse of the right of the social body to ensure, maintain, or develop its life" (136). Thus, bringing death to some becomes, paradoxically, a way "to ensure, maintain, or develop" life. The power to protect life from threat migrates as well, becoming one "that endeavors to administer, optimize, and multiply it, subjecting it to precise controls and comprehensive regulations" (137). We see this regulative practice in Pasteurian procedures related to food safety, which administer and optimize (human) life by killing (microbial) life. Slonczewski's figuration of microbial communities with distinct cultures living within and interacting with their human hosts adds new intensity to Foucault's description of the genocidal quality of this new biopower that disperses death to foster life: "Wars are no longer waged in the name of a sovereign who must be defended; they are waged on behalf of the existence of everyone; entire populations are mobilised for the purpose of wholesale slaughter in the name of life necessity: massacres have become vital" (137).

Brain Plague

The final novel in the Elysium cycle, ***Brain Plague*** (2000), imagines the future that emerged from the concluding events in ***The Children Star***, after humans learned to converse with intelligent microbes inside them and some of these microbe cultures went out into the larger universe with their human hosts. *Brain Plague* takes place mainly on Valedon, where artist Chrys decides to allow herself to become infected by sentient microbes, who are now a part of the Free Fold collective (although they need a human host within which to reside). Chrys lives in a city that is starkly divided between spaces of privilege and conspicuous consumption and those of poverty and despair. Medical services are starkly divided between the opulence of Plan Ten and the overburdened resources of Plan One, leading to vast inequities of lifespan based on economic resources. Slonczewski's vision of microbial posthuman ethics responds to these limitations of a hierarchical culture that humans alone have made.

Chrys is drawn to using these microbes, called "brain enhancers," for financial reasons but also because she hopes they will help with her art. They are promoted as parallel microprocessors through which one can augment one's brain, but citizens are also aware of the nefarious "plague" micros that infect human hosts and turn them into desolate shells of their former selves, living only to serve the needs of these micros, under the direction of a leader called Endless Light. When Chrys learns that the brain enhancers and the plague are genetically the same species, she almost withdraws her consent, but a friend convinces her that the sentient microbes need to be understood based on their distinct values not only via their biology: "these are a completely different culture. Entirely different history and lifestyle" (28). The novel mirrors issues about good governance and collectivity on micro- and macro-scales. The same gaps between the wealthy and the impoverished that Chrys observes in the world around her are mirrored in the micro-community that develops within her body. Chrys critiques, for example, the outbreaks of xenophobic hatred directed against intelligent machines and immigrant "simians"—those with "*Homo gorilla* ancestry" (13) on her way to the lab, but later expresses her own instinctive fear that the micros are "taking over," before being chastised, "Iridians always say that, about the latest new immigrants" (25). Much later, the micros within Chrys have to confront their own systems of governance and resource distribution as the colony grows large and falls prey to uneven development: "In the old days, everyone shared alike; but now, as their world neared a million strong, some, like Jonquil [one of the individualized micros within Chrys], grew rich enough to spend all their palladium in the nightclubs, whereas others floated by with nothing" (117).

Chrys and her micros learn to communicate and to create a mutual culture in which each can flourish through an entwined "care of the self" similar to the interactions Paxson observes around cheese production. Chrys must consume an arsenic supplement to provide needed minerals to the micros, who evolved on a world rich in this mineral, while the micros ensure that this poison does not harm Chrys and also calibrate other aspects of her physiology toward maximum health. Chrys's micros, the Eleutherians, develop a distinct cultural identity that is a reflection both of their internal generational cultures and their interactions with Chrys, whom they regard as both world and god. The rapid changeover in micro-generations means that Eleutheria undergoes many generational eras, including a period in which young rebels, rejecting religion, refuse to communicate with Chrys at all. Similarly, Chrys learns to accommodate herself to the rapid pace of Eleutherian life, to how quickly generations change and the individuals with whom she converses are replaced by others. She also learns that care for herself is also care for a community: In one incident, she fails to heed their warnings about treating a cut to minimize blood loss and a key micro-community member is "lost in that rush of blood from her arm. Mopped up and gone forever" (237).

Eleutheria evolves into a culture predicated upon mutuality, empathy, and compassion, all qualities important to Chrys and to her artistic vision. Chrys and the Eleutherians offer an ideal of the kind of hybrid mutuality of a posthumanist identity grasped of as the supra-individual: Her art is changed by the influence of the micros on her physiology of perception. While she is anxious at first whether this new art is truly "hers," she eventually comes to terms with her role in this creative process when she stops thinking of herself narrowly as the "I" of liberal humanist autonomy. Embracing a more inclusive sense of self, a posthuman self that includes Chrys, Eleutherian, and—crucially—the third identity and vision that they produce through their interactions, Chrys and the Eleutherians begin to produce both art and architecture that exceeds what either was capable of alone.

Their work is committed to principles of social justice, particularly the architectural work that seeks to make buildings both beautiful and inhabitable, in response to their horror about an elite culture of commodity architecture that creates beautiful structures that tower over the hundreds of homeless in the streets. Chrys comes to describe the art that she produced before her collaboration with the Eleutherians as "wilderness without people" (105); with them she is able to create something new. Chrys and her micros can be understood as akin to the relation between artisanal cheesemakers and the microbial cultures that have evolved in specific established cheese houses that have been in operation in, for example, France, for decades, even centuries, sustained by ongoing craft practices: "natural flora and fauna, they materialize as specific communities within ecologies of human practice. To speak doubly of cheese cultures—bacterial and human—is thus no idle pun" (Paxson 2008, 25).

The collaborative model of posthuman identity that Chrys and Eleutheria embody is contrasted by two negative poles in *Brain Plague*, one in which the humans wage war on the existence of the micros in the name of "defending" (in Foucauldian terms) human life, and the other where the micros reduce humans to dopamine-addicted slaves in a similar defense of their own lives. Selenite, a more corporate artist than Chrys and at times Chrys's patron, exemplifies one side of this dichotomy. Rather than allowing her micros a degree of autonomy and thus creating a space for something new and hybrid to emerge between them, Selenite and those like her carefully control the breeding of resident micros: "Some of us select which offspring to merge. We cultivate our strains for essential skills, while discarding less helpful traits. In the end, they'll merge with our own brains—true brain extensions" (129). For this human faction, micros are mere tools, not fellow beings, extensions of a narrow set of "essential skills" rather than extensions and transformations of the self into the collective supra-individual. On the opposite side of this binary are the micros of Endless Light, led by their leader Rose, who view humans as merely a world of resources to be harvested and controlled. Endless Light communities lack the mutuality of supra-individuals like the community Chrys cultivates: Their culture seeks to protect only its own micro-identity regardless of which host they migrate to, and they control the host through dopamine, forcing him or her to consume only the resources the micros need, until that body is used up and they move on to another.

Just as Rod in *The Children Star* was protected from infection by pathogenic micros through the efforts of his resident colony, so do micros that are allowed to develop autonomous cultures protect their host. Thus, the best protection against "plague" micros is not the thanatopolitical destruction of all microbial life—or the milder version of keeping one's micros subservient—but instead it is a mutuality between host and sentient micros. The novel ends with a description of Eleutheria as "no genetic race, nor a physical place, but a way of being, a path of endless life" (303), a commitment to social justice that Chrys and the Eleutherians express in a pledge to use their skills to build public housing. Endless life, Chrys's way, in contrast to the Endless Light of the plague micros, endorses an openness to mutation and change, to mutuality and a self that is larger than the individual, a self that is collective and always-in-process. Endless life emerges from values aligned with Tsing's collaborative survival and rejects a biopolitics that fosters life through rigid control that "wages war" in the name of protecting this life.

The Highest Frontier and the Latent Commons

Tsing (2012) concludes with what she calls the latent commons of multispecies entanglement and compromise, rather than the idealized vision of conflict-free mutuality that she dubs the utopian commons. The idea of latent commons implies an understanding of the resources of the world held in common and shared across species, not merely a commons among and for humans. This posthuman perspective on the commons requires us to acknowledge that simple harmony is not achievable, that life across ecosystems necessarily involves relations of predation and tensions among divergent ecologies: "Latent commons are those mutualist and nonantagonistic entanglements found within the play of this confusion. Latent commons are not good for everyone. Every instance of collaboration makes room for some and leaves out others. Whole species lose out in some collaborations. The best we can do is to aim for 'good-enough' worlds, where 'good-enough' is always imperfect and under revision" (255). Set, unlike her Elysium novels, on and above a near-future planet Earth, Slonczewski's most recent novel involving microbial life, *The Highest Frontier* (2011), offers a potential vision of the difficulties of trying to enact such latent commons.

Much of the story in *The Highest Frontier* takes place a hundred years from now on an orbital "spacehab." At least in such manufactured circumstances, humans seem to have perfected their control over the biological world. The station is home to a university, Frontera, mainly for children of the privileged, a safe haven for them given that Earth has been colonized by an alien species, Ultraphytes, single-celled organisms, like microbes, but which grow to a perceptible size. Ultraphytes are dangerous because their different biology means that they release cyanide when under stress, and thus they can be poisonous to earth lifeforms. Ultraphytes are treated as a public hazard and never engaged with as anything but a toxin that must be contained and destroyed in what is described as the "War on Ultra" (10). Ultraphytes also seem to be the last site of the natural world retaining agency vis-à-vis human culture, since all else has been safely transformed into tool for human use, especially within the environment of Frontera where only manufactured species are permitted. For example, HIV is now an acronym for "human improvement vector," the original virus "tamed" to become a monitoring system for human health (22). Most objects and habitats are grown from amyloid, "a bacterial protein that self-assembled any form" (15)—including foods. The main method of manufacture is 3D printing, and anything short of a living organism, including viruses which are at the cusp between living and inert, can be printed.

Frontera pursues an ideal of human control over all other species. As a college administrator extols in promotional tones, their orbital campus is "home to birds and deer, free of mosquitoes and yellow plague, untouched by drought or flood. ... Indeed, all of Earth's rivers could dry out, and all of Earth's oceans could swallow her shores, while our pristine habitat of Frontera would remain untouched" (40). Thus, while most of those at Frontera adhere to ecological values, the environment they have created also allows for an ideology to take root in which some people are unconcerned about the ongoing destruction of Earth's ecosystems for the sake of profit, since life in space can provide a substitute, at least for a privileged few. The animal species that populate Frontera are only those desired by humans, and many are engineered to be miniaturized versions of their originals, such as elephants or bear, to better suit the compact habitat but also to better serve as tools for human needs. Although Frontera is understood to be a perfected environment by its inhabitants, the degree of their distance from other species is somewhat disconcerting, as in the example of when our protagonist, Jenny, touches the trunk of a miniature elephant: "Jenny tugged at the trunk, disgusted by its squishy feel. She had never had live pets; she'd been kept away from animals all her life. In her toybox,[6] she blinked her 'safe' button. The elephant squealed in pain at the mild electric shock" (60). Frontera does not harm any animals to produce food or anything else, as we are repeatedly reminded, but the lack of relations of exploitation has become a lack of any kind of relations at all.

Three plots intersect in the novel. First, a scientist's study of plant communication systems, via pheromone-like secretions, to explore a kind of distributed rather than centralized thinking and thus cultivate a quality the novel calls wisdom. Second, the religious fundamentalist Centrist party, which advocates a return to conservative values while valorizing profit, and the Unity party, that focuses more on social-service provision and slowing climate change, are waging an election campaign for the next president of the United States. And finally, the human relationship with the ultraphytes develops an embodied spokesperson in the figure of Mary. A schoolmate of Jenny's, Mary initially appears to be a human girl whose chronic illnesses require her to use a largely prosthetic body. These storylines all intersect in a focus on biology understood as a system of cooperation rather than one of competition. I will focus on the third storyline about the human relationship to the "invading" ultraphytes.

About mid-way through the novel, humans discover that many creatures they thought of as purely earth-based already have sequences of ultraphyte DNA and, as they explain, the integration of this ultraphyte genomic sequence into the genome of the earth-based species means "RNA conversion to DNA by an ultra-encoded ribonucleic acid transfer-integrase," or, in more colloquial terms, that "ultraphytes can turn Earth creatures into ultra" (266). Those familiar with sf might recognize here an allusion to John Campbell's 1938 novella *Who Goes There?* that was first adapted in 1951 as *The Thing from Another World* and again in 1982 by John Carpenter as *The Thing*. In this much-loved story, an alien invader is able perfectly to imitate any earth-based lifeform, threatening to thus take over the entire planet by replacing all life with alien life. This produces a kind of ontological panic among the men staffing the Antarctic research station that makes this discovery, until they are finally able to develop a test that can discern "real" humans from alien imposters, thereby destroying the invader and indeed any life through which it might escape Antarctic isolation and contaminate the rest of the planet. A vehemently xenophobic tale of human exceptionalism, *Who Goes There?* became a classic of early sf.

The Highest Frontier has a radically different response to this scenario of invasion, one that emphasizes mutuality and hybrid subjectivity rather than panicked purification. When ultraphyte DNA is discovered in other species, Jenny realizes that it is possible that ultraphyte genes exist in humans as well, especially given that "human genomes were already full of parasitic DNA" (180), an aspect of the human as supra-individual discussed above. The mainstream political response to this knowledge is panic like that depicted by Campbell, a massive and violent destruction of any life suspected of ultraphyte contamination, which on Frontera includes not only the ultra samples in their research laboratory but also any other research subjects: "Besides the ultras, they'd burned all the rats and chickens too. Jenny felt sick" (387). Jenny, however, sees another way that opens the door for collaborative survival with the ultraphytes, a conclusion she reaches when she realizes that her classmate, Mary, is itself an ultraphyte.

As Mary demonstrates, ultraphytes are not malevolent agents bent on the destruction of human life, but simply a different kind of life whose biology is not immediately compatible with human biology. The ultraphyte presence calls out protocols for co-existence, for newly imagining the world as a latent commons. As Mary points out to her classmates in an early scene, before they realize her identity, the characterization of ultras as toxic emerges only from the point of view of life based on a different kind of biology, and one does not have to respond to this difference by demonizing it. "Humans make poisons too," Mary calmly points out as her classmates articulate their fear of ultraphytes, "and humans are beautiful" (48). Mary is a community organism, an ultraphyte supra-individual, that is, many ultraphyte cells linked together to form the collective entity that operates the prosthetic human body and thus blends in with the humans. In this body, ultraphytes study humans just as humans study them. It is notable, however, that Mary's protocols for study are non-lethal and non-invasive, immersing herself in human community rather than isolating humans in a laboratory and subjecting them to controlled tests. Mary also offers them another way to think about subjectivity as collective and distributed, a model of the posthuman. Her individual ultraphyte cells are "more like 'citizens' of a colony" than the constituent parts of her single identity, and decisions about action are made by the cumulative choices of each cell: "the whole group takes a vote, a hundred times a second" (198).

Mary models a kind of agency that can think multiple things at once but still come to a collective decision and act. This ideal is paralleled in the two other main plot lines, about plant research and about the election. Guided by their teacher's love of plant thinking, via a distributed central nervous system rather than a single locus of thought in the brain, Jenny and Mary modify plants to interact with the human nervous system and secrete a chemical signal for this capacity for multi-perspective thinking, which they call wisdom. In an experiment that goes somewhat awry, these plants influence the rhetoric at a debate between Centrist and Unity candidates, compelling both to move away from the more extreme poles of their rhetorical positions and find a collaborative middle ground that can provide the foundation for a sustainable political system and so achieve positive change rather than remaining mired in deadlock. This archetype of a more beneficial political order emerges from a commitment to collaborative survival rather than to zero-sum games of singular victory. The novel, the first in a planned trilogy, ends with most of humanity still afraid of ultraphytes and pursuing Mary as a threat. Nevertheless, the protagonist Jenny has extended this lesson of mutuality toward the future of human/ultraphyte relations; I anticipate the works to follow will build on her example.

In conclusion, Slonczewski's novels of microbial life offer a posthuman ethics of multispecies collaborative survival, a perspective that begins with the recognition that the human itself is colonized by—or, perhaps better, composed via—multiple microbial species within and without. Avoiding a naïve vitalism that would simply collapse all boundaries between species or that would disavow the inevitable conflict among them, this microbial posthumanism instead points toward the different kinds of social relations that are required to embrace the collaborative survival upheld by Tsing. Tsing argues that this stance must materialize from within an acknowledgment of the precarity of all life, which requires being vulnerable to and through others. The human that participates in the ethics of collaborative survival becomes something else, becomes posthuman: "Unpredictable encounters transform us; we are not in control, even of ourselves. Unable to rely on a stable structure of community, we are thrown into shifting assemblages, which remake us as well as our others" (Tsing 2015, 20). Collaborative survival changes us, biologically and socially, and Slonzewski's work offers a concrete depiction of how these levels are co-produced. Her fictions unite rigorous accounts of the microbiology that underpins her human/micro-collaborations with wide concerns for the larger social worlds in which her stories are set. Issues of social justice—poverty, racism, environmental exploitation—are key concerns in these worlds, and the way toward ameliorating them consistently rises out of rethinking the self as a human/micro-supra-individual.

The ideals of microbial posthumanism embodied in these fictions teach us that "*human nature is an interspecies relationship*":

> *Human exceptionalism blinds us.* Science has inherited stories about human mastery from the great monotheistic religions. These stories fuel assumptions about human autonomy, and they direct questions to the human

> control of nature, on the one hand, or human impact on nature, on the other, rather than to species interdependence. One of the many limitations of this heritage is that it has directed us to imagine human species being, that is, the practices of being a species, as autonomously self-maintaining. (Tsing 2012, 144)

Slonczewski's microbial fiction teaches us that humans are anything but "autonomously self-maintaining"; we are always already involved in collaborative, multispecies modes of posthuman existence that make the world and ourselves through the social relations we foster in these encounters. Her work calls on us to recognize that biopolitical "care of the self" is necessarily also care of other species, and that from this mutual care flow the kinds of social structures that can sustain life for all species. The novels provide concrete examples of what Paxson calls an "entirely different set of social relations [that] can be unpacked" (40) in new ways of thinking of humans and microbes as co-producers of the world.[7] In Rod's commitment to a model of migration and settlement there is more than destructive resource extraction in *The Children Star*. The public housing project that will sustain Chrys and the Eleutherians artistically and their community materially in *Brain Plague* forecasts the more collaborative human politics and anticipated future of human/ultraphyte relations anticipated at the end of *The Highest Frontier*. Entire new sets of social, economic, even political arrangements—new ways of living—can flow from reconfiguring human/microbial relations. Slonczewski's fiction offers us a posthuman ethics whose transformations aspire far beyond the mere augmented bodies of her characters.

Notes

1. Research on the microbiome is changing our ideas about health and disease. Among the paradigm shifts inaugurated by this research is a move away from studying individual species of microbes and a move toward conceptualizing them as a community in continual exchange and interaction. The object of study becomes ongoing process rather than a fixed object. This shift emerges from evolutionary biologist Lynn Margulis's own work as well as her contribution to the influential Gaia hypothesis, developed with James Lovelock, that understands the planet as an entwined system that interrelates organic and inorganic, organism and environment, into a complex whole that can be conceptualized as an autopoietic system. Clarke (2017) argues that the increasingly centrality of symbiosis to recent biological

theory can be greatly if not entirely credited to Margulis. Developing a Gaian theory of the biopolitical, he notes that Margulis's preferred term for what I am calling the supra-individual is the *holobiont*, as produced by symbiogenesis, "the development of new life forms by the incorporation or colonization of one or more organisms into or by one another" (204).

2. Paxson (2014) cautions against letting the pendulum swing too far in the other direction and thus simply embrace all microbes as benign, even as allies. She resists some of the ways that her phrase "post-Pasteurian cultures" has been taken up into an ethos that refuses all cautions against microbial contact. Some microbes are very dangerous pathogens; others are centuries-old "collaborators" whose unique history produces things such as beloved varieties of cheese from long-established French dairies, for example. The issue is understanding the variety of microbial life and thinking through a politics of our mutual co-existence with it.
3. There is a body of literature about communication within microbial communities, called "quorum sensing," which is spoken of as a kind of chemical language. See, for example, Czárán and Hoekstra (2009). There is also a body of work resistant to the idea of calling this communication or a language. See, for example, Diggle et al. (2007).
4. It may well be, as Landecker (2016) suggests, that humans are already communicating with microbes in such aggressive terms on a regular basis. She points out that the substances we call antibiotics are compounds mined from microbes themselves, substances that are found in "sub-inhibitory" concentrations in the soil, and thus might be best understood as "means of communication and gene regulation in and between bacterial species. … If antibiotic molecules are actually not things that kill, but things that communicate and coerce, their use has been at deafening concentrations for many years now" (27).
5. The all-capital typeface is used to denote the content of what microbes communicate to their host via flashing letters, in this case to Rod.
6. The toybox is a sort of virtual reality and social media augmentation that simultaneously allows people to immerse themselves in virtual spaces and augments their perceptions of material reality with data overlays. Slonczewski describes the various "windows" of the toybox as "a cube of light that hovered just before [one's] eyes" (2011, 9).
7. As Paxson's compelling work shows, such possibilities are more than fictional. For her, artisanal cheese production is the model of a different way for humans to live collaboratively with microbes, whose contributions to the process are acknowledged and valued, contra the Pasteurian culture of mass sterilisation. Producing cheese in this way changes more than just the cheese: "raw-milk cheese—produced safely only at an artisan scale—[can] provide a future for family farmers and … preserve Vermont's 'working

landscape' because it requires clean milk from animals on pasture and fresh-dried hay, not commodity corn" (2008, 40).

References

Braidotti, Rosi. 2013. *The Posthuman*. Cambridge: Polity.

Clarke, Bruce. 2017. Planetary Immunity: Biopolitics, Gaia Theory, the Holobiont, and the Systems Counterculture. *General Ecology: The New Ecological Paradigm*, ed. Erich Hörl with James Burton, 193–215. London: Bloomsbury.

Czárán, T., and R.F. Hoekstra. 2009. Microbial Communication, Cooperation and Cheating: Quorum Sensing Drives the Evolution of Cooperation in Bacteria. *Public Library of Science One* 4 (8). Online.

Diggle, Stephen P., Andy Gardner, Stuart A. West, and Ashleigh S. Griffin. 2007. Evolutionary Theory of Bacteria Quorum Sensing: When Is a Signal Not a Signal? *Philosophical Transactions of the Royal Society of London Series B Biological Sciences* 362 (1483) (July 29): 1241–49.

Foucault, Michel. 1978. *The History of Sexuality. Volume 1: An Introduction.* New York: Vintage.

Haraway, Donna J. 1991. "A Cyborg Manifesto: Science, Technology, and Socialist-Feminism in the Late Twentieth Century." *Simians, Cyborgs, and Women: The Reinvention of Nature*, 149–81. New York: Routledge.

———. 2007. *When Species Meet*. Minneapolis: University of Minnesota Press.

Hird, Myra J. 2009. *The Origins of Sociable Life: Evolution After Science Studies.* Houndmills, Basingstoke: Palgrave Macmillan.

Landecker, Hannah. 2016. Antibiotic Resistance and the Biology of History. *Body and Society* 22 (4): 19–52.

Latour, Bruno. 1993. *The Pasteurization of France*, trans. Alan Sheridan and John Law. Harvard: Harvard University Press.

Paxson, Heather. 2008. Post-Pasteurian Cultures: The Microbiopolitics of Raw-Milk Cheese in the United States. *Cultural Anthropology* 23 (1): 15–47.

———. 2014. Interlude: Microbiopolitics. In *The Multispecies Salon*, ed. Eben Kirksey. Duke: Duke University Press. Kindle.

Shapiro, James A. 2015. Bringing Cell Action into Evolution. In *Earth, Life, and System: Evolution and Ecology on a Gaian Planet*, ed. Bruce Clarke, 175–202. New York: Fordham University Press.

Slonczewski, Joan. 1998. *The Children Star*. Rockville, MD: Phoenix Pick, 2010.

———. 2000. *Brain Plague*. Rockville, MD: Phoenix Pick, 2010.

———. 2011. *The Highest Frontier*. New York: TOR.

Tsing, Anna. 2012. Unruly Edges: Mushrooms as Companion Species. *Environmental Humanities* 1: 141–54.

———. 2015. *The Mushroom at the End of the World: On the Possibility of Life in Capitalist Ruins*. Princeton: Princeton University Press.
Wolfe, Cary. 2009. *What Is Posthumanism?* Minneapolis: University of Minnesota Press.

CHAPTER 7

The Future at Stake: Modes of Speculation in *The Highest Frontier* and *Microbiology: An Evolving Science*

Colin Milburn

For the microbiologist and science fiction writer Joan Slonczewski, speculation plays a vital role in the enterprise of science and, indeed, our entire high-tech culture. How could it be otherwise? But the question is less how to speculate knowledgably, informed by the best empirical data and laboratory research, than how to speculate responsibly—especially at a moment when the advancement of science and technology seems to be propelled by venture capital and futures markets, forward-looking promises and dreams of disruptive innovation.[1] It is a question of how to speculate with wisdom, at a time when the shape of things to come seems to be framed by competing visions of profitability versus catastrophe—the antinomies of the futures industry—with no clear way to navigate between, much less chart an entirely different trajectory.[2]

Traversing her occupations as a scientist, a novelist, and a professor at Kenyon College, Slonczewski has affirmatively positioned science fiction

C. Milburn (✉)
Departments of English, Science and Technology Studies, and Cinema and Digital Media, University of California–Davis, Davis, CA, USA

© The Author(s) 2020

B. Clarke (ed.), *Posthuman Biopolitics*,
Palgrave Studies in Science and Popular Culture,
https://doi.org/10.1007/978-3-030-36486-1_7

as a way to address these questions, a device for tuning the imagination and attuning our society to the politics of speculation. In her work—her scientific writings, her provocative novels, her pedagogical methods—Slonczewski invests science fiction with an educational responsibility, not only modeling different ways of thinking about the future of science, but also critically scrutinizing the cultural practices of speculation that aspire to shape and contain this future in advance. In other words, Slonczewski casts science fiction as a self-conscious, recursive discourse on the calculus of anticipation: a way to speculate on speculation itself.[3] This chapter looks at two examples of Slonczewski's work that underscore the self-reflexive, pedagogical affordances of science fiction: her novel *The Highest Frontier*, which depicts the values of a liberal science education in a society of rampant speculation, and her textbook *Microbiology: An Evolving Science*, co-authored with John Foster, which puts these values into practice. Each of these texts suggests that speculation is a double-edged sword, describing both the future-generating and future-confining forces of our world. But the virtue of science fiction is that it can teach us to see the difference and imagine better.

To Space—And Beyond

The future is built on speculation—this much is clear in *The Highest Frontier* (2011). It traces the political conditions of the United States in the next century by focusing on the adventures of a student named Jenny Ramos Kennedy in her first year at Frontera College. Frontera is located in an orbital space habitat, representing the outer frontier of human expansionism. The imagery of the frontier—the edge of territory and knowledge—pervades virtually every aspect of social, religious, and scientific life in this future, recalling aspirations that have characterized the American project throughout its history. Indeed, the civic discourse in the narrative persistently, obsessively references Frederick Jackson Turner's frontier thesis, the figure of Teddy Roosevelt as visionary frontiersman, and the iconography of manifest destiny.[4] Slonczewski shows how the symbol of the frontier—orienting patriotic ambitions and technoscientific investments—is a function of the political economy of speculation, the cultural manufacturing of things to come.

For one thing, the society of *The Highest Frontier* is indebted to science fiction. Any number of its innovations derive from the pages of twentieth-century novels. The popular sport of "slanball," for instance, played in low-gravity environments using technologies of psychokinesis, pays homage to A. E. Van Vogt's 1946 novel *Slan*. While the posthuman slans of Van Vogt's novel possessed innate psychic powers, the developers of slanball in Slonczewski's text have instead invented a practical workaround for everyday psionics—a technoscientific modification of the fabulous trope. Similarly, the political campaigns that seem to dominate American consciousness in *The Highest Frontier* rely on predictive big-data analytics that trace their origins back to Isaac Asimov's *Foundation* novels, as Jenny notes when she visits a museum: "On the wall an old poster of Hari Seldon, the fictional psycho predictor; every campaign had their own 'Seldon' now" (258). In Asimov's novels, Hari Seldon's science of psychohistory allowed him to foresee—and to some degree engineer—the future of human civilization by revealing how apparently unpredictable social phenomena might look more like certainties at higher scales. Yet Seldon's system actually assumed that nothing is truly inevitable, merely more or less probable—which is why his speculative enterprise relied on some colossal insurance plans, hedging against unknowable risks. To be sure, in *The Highest Frontier*, the anticipatory calculus of risk analysis and futures markets has become so fundamental to the social order that even things that once seemed inevitable, such as the old certainties of death and taxes, have now been rendered entirely probabilistic. People casually reengineer their chromosomes to increase the likelihood of longevity, and the US taxpayer system has been replaced by the tax*player* system, where citizens are enabled—or rather, required by law—to gamble their taxable earnings in casinos, placing bets, playing games of chance with their anticipated debts to the state. The entire fiscal infrastructure of the country is now based explicitly on wagers, scrutinized with Seldonian modes of statistical analysis that promise the effective management of risk.

However, there are actually two forms of speculation at stake in *The Highest Frontier*. One form of speculation considers the future in terms of the present, modulating the future with tools for mitigating and profiting from risk, including stochastic models, insurance schemes, financial derivatives, industry roadmaps, and technological forecasts. But there is also a form of speculation that opens to uncertainty, envisioning the conditions of the future as less knowable than transformable—allowing for other worlds than this one. In their book *Speculate This!* (2013), the anonymous scholars of the "uncertain commons" distinguish between these two forms of speculation, describing the first as *firmative*:

> This specifically modern form is what we call *firmative speculation*, a firming (from the late Latin *firmā-*) or solidifying of the possibilities of the future. It is a speculative mode that seeks to pin down, delimit, constrain, and enclose—to make things definitive, firm. The ur-image of such agency is the *firm*, a type of business house (emerging in Germany in 1744) that capitalizes on market conditions ... Firms draw on expertise in speculative science materialized in risk instruments such as insurance, annuities, and stock options. These instruments render firm the uncertain future, enclosing us within a relatively secure horizon—a firmament, as it were, seemingly fixed over the earth. (ch. 1)

Yet this is not everything, the be-all and end-all. At least, we can imagine otherwise. The uncertain commons describes another form of speculation as *affirmative*:

> On the other hand, there is expectation, conjecture, and anticipation: modes of living that recognize the dormant energies of the quotidian and eventualities that escape the imagination. We call these modes *affirmative speculation*. To speculate affirmatively is to produce futures while refusing the foreclosure of potentialities, to hold on to the spectrum of possibilities while remaining open to multiple futures whose context of actualization can never be fully anticipated.... In this sense affirmative speculation affords modes of living that creatively engage *uncertainty*. Its stakes are resolutely collective: often sabotaging individuated and privatized prescriptions, it builds on the tentative mutualities that arise in the face of uncertainties. In short, affirmative speculation embraces ways of *living in common*. (ch. 1)

In *The Highest Frontier*, these speculative alternatives are embodied in the two American political parties of the future, manifested in their policy platforms and their relations to the horizon of science and knowledge. The Centrist Party upholds a fundamentalist belief that the visible night sky represents "the Firmament," both a physical limit and a destiny: "'Firmament' was the Centrist word for the sphere of the biblical heavens that centered on Earth" (14). Faced with the realities of ongoing climate change on Earth, the escalation of inhospitable conditions, the Centrists nevertheless insist that existing human practices cannot or ought not be changed. Instead, they envision a future where human survival depends on the development of more orbital habitats—little heavenly retreats fixed securely in the grip of the Firmament—allowing well-to-do families to abandon the planet and any lingering sense of responsibility. It is a speculative vision of catastrophe, an eschatology that becomes a self-fulfilling prophecy.

On the other hand, the Unity Party—while taking a "big tent" position and tolerating diverse beliefs ("If your world has a Firmament, we can still work together" [373])—maintains that the Earth is not the center of the universe. The Unity platform supports further exploration of outer space, including a mission to Jupiter, advocating the expansion of thought and curiosity. Yet even while projecting outward, the Unity position reflects back on itself to address extant conditions on Earth, the deeper causes of global climate change. (This also includes reassessing some practices formerly considered green and progressive, such as the proliferation of terrestrial solar arrays, that have ironically extended problems in other directions due to insufficient foresight.) Unity speculates on a future that diverges from the present, pressing beyond the Firmament and the firmative as such.

During Jenny's first semester at Frontera College, the tension between these two modes of speculation reaches a climax due to preemptive actions taken by the incumbent Centrist government. An alien life form has invaded Earth: ultraphytes, an ultraviolet-photosynthesizing quasi-species capable of rapid mutation, lateral genetic transfers, and adaptive evolution. Despite efforts to curb their spread, ultraphytes have been migrating everywhere and swapping genes with native Earth species. The ensuing "War on Ultra" lays bare the logics of speculative securitization, showing how preemptive strikes often end up producing the very futures they had sought to eradicate (cf. de Goede 2012; Parisi and Goodman 2005). In fear that weaponized ultra might appear at some point in the future, the Centrist vice president—colloquially called the "Creep"—authorizes a secret military project to weaponize ultra in advance: "As a defense experiment. That is always the excuse to make biological weapons: to devise a defense." Jenny learns of the analogous history of anthrax: "That's why the government made weapons-grade anthrax.... To test defenses" (389). The Creep's scientific team thus creates a threat that had not previously existed. They assemble a sentient, humanoid colony of ultraphytes that poses as a student named Mary Dyer and enroll her at Frontera College. While the Creep's plan uses Frontera as a testing ground to study weaponized ultra, the space college provides an environment where the

Mary colony develops in unexpected ways, especially through her interactions with professors and fellow students. (Serendipitous learning and intellectual growth are, of course, notorious effects of liberal education.)

From its first sentence onward, the novel thematizes such unexpected twists and turns of futurity: "The space lift rose from the Pacific, climbing the cords of anthrax bacteria" (9). The language of science fiction, requiring a semantic reorientation—the generative potential of the "as if" and the "yet to be," the science-fictional subjunctive—opens the future to difference (cf. Delany 1977; Saler 2012). It casts forth a time to come when even dread diseases and secret military weapons might defy expectations and become allies, transportation infrastructures, everyday commodities:

> Jenny rose in the lift climbing the anthrax bacteria from Earth to orbit. The bacteria with their nanotube cell walls had first been engineered at Fort Dietrich to test defenses against weapons-grade anthrax. Now the anthrax grew kilometers long out in space. Self-healing cables were just the thing to hoist freight up through Kessler debris, for growing cells soon replaced those severed. Jenny's mother had managed the Anthradyne IPO, the first anthrax farm in space. (16)

As a fortuitous spinoff of military research, the nano-enhanced anthrax now assists humans to climb into orbit. Once the bane of cattle and ranchers, the nightmare of military strategists who dreaded its weaponization—and who then perplexingly developed it as a weapon to preempt this threat—anthrax has since become a tool for opening the future.[5] Borne forth on the dynamic looping of firmative and affirmative speculation, these microbial cables sketch a trajectory beyond Earth, beyond the present. And yet, while plotting lines of flight outward, the anthrax technologies are recaptured by mechanisms of finance capital that parcel up the future and delimit its options under the sign of the IPO and the market. Indeed, Jenny's mother advises an investor, "Yes, *hombre*, anthrax futures are just the thing" (10). The futures of anthrax are rendered yet one more tradable thing—"just the thing"—another object of "the mode of firmative speculation that *produces* potentialities and then *exploits* and thus *forecloses* them" (uncertain commons 2013, ch. 2).

The novel charts how firmative agendas actively produce future-opening effects, and vice versa. But at the same time, the mechanisms of containment are incomplete—no matter how unlikely it may seem, the future could always go astray. For example, the Mary ultra makes a crucial scientific breakthrough, bioengineering a new species of plant that releases semiochemicals affecting the human capacity for self-reflection and contemplation. That is to say, she creates *wisdom plants*. The wisdom plants represent a line of scientific research previously considered infeasible, unrealistic. Anouk, one of the other students in Jenny's biology class, notes that their professor's neuro-botanical research on the chemistry of wisdom had never borne fruit: "But never wisdom plants. Over and again, students have assayed different combinations of genes, testing different neural networks for 'wisdom.' Still nothing" (332). She sums up the prevailing consensus: "There is no such thing as a reverse control for wisdom." But Mary calmly replies, "There is now" (333). In the novel, then, the wisdom plants are science fiction become science fact. They conjure forth the vicissitudes of the future even in the context of an overextended present.

Serendipitously, accidentally, the wisdom plants wind up on a Frontera stage where the Unity and Centrist presidential candidates have agreed to conduct a televised debate. The strange semiochemicals the wisdom plants release into the air create the conditions for compromise between the candidates. This leads to a Unity win in the presidential election and prepares the groundwork for change, both a restructuring of unsustainable conditions on Earth and a rejection of closed-world ideology, the politics of relentless inwardness (cf. Edwards 1996). As a result, some characters note that the wisdom plants, emitting subtle vapors that diffuse across barriers of ideology, seem to embody the function of the university as such: "And some say that wisdom breathes in the very air of Frontera." In this regard, the trope of "the highest frontier" now also discloses another meaning. As the president of Frontera College says, "Wisdom is the highest frontier" (393).

The novel concludes with an emergent, ambiguous friendship between Jenny and the remains of the Mary entity, whose ultraphyte cells have now been dispersed into the intimate structures of the space hub. The future is left open for adventures and sequels yet to come. But this is now a future where wisdom plants are known to exist, with all that they signify. It is a future where alien microbes are rapidly evolving in relation to the creatures of Earth, experimenting with new ways to flourish in common: "Who knows what ultra would come up with next?" (443). It reinforces the theme presented in the novel's opening sentence, projected by the twists of science-fictional language—namely, the potential of the future to defy expectations, to deviate from roadmaps and investment strategies, security lockdowns and technology forecasts—reminding us that such potential is already latent in the present.

In focusing on the limitations of a firmative speculation—a mode of speculation that manages the future as risk and contracts possibilities in terms of the present—versus an affirmative speculation that opens to uncertainty, *The Highest Frontier* highlights the epistemic functions of science fiction as a narrative genre. Of course, as Slonczewski reminds us, science fiction can be just as prone to fantasies of firmative speculation—recall Asimov's Hari Seldon—as to explorations of difference. Yet by narrating the production of futurity, thematizing the processes of innovation and change, science fiction constitutes *a second-order discourse on speculation.* It recursively speculates on the speculative mechanisms of our own world, the oscillations of future-closing and future-opening trajectories that shape science and society.

By showing how the future gets framed by the firmative modes of the present, science fiction catalyzes a thinking otherwise. For regardless of whatever contents it might feature, science fiction is one of the few cultural forms that, as Fredric Jameson has argued, enable us to "think the break" (232). Or, as Slonczewski has suggested, it conjures "a way out": "My science fiction offers way out—a way forward to say, 'We can do something for our planet'" (qtd. in Jones 2011, 59). Precisely by drawing attention to the difficulty if not impossibility of envisioning a future radically distinct from the contemporary and the expected—that is to say, beyond the limits of the Firmament—science fiction provokes ruptures from the opposite direction, spaces for critical attention, altered perspectives on the social forces that shape all practices of speculation. It elicits reappraisal of our own limitations as organisms on a finite Earth, the ties that bind our forward-looking ventures to this world—indicating perhaps that the highest frontier is less something "out there" than already "in here," a capacity for wisdom yet undiscovered.

In Science—and Out

More than a decade separated the publication of Slonczewski's previous novel, *Brain Plague* (2000a), and *The Highest Frontier*. In the intervening years, while developing the story of *The Highest Frontier*, Slonczewski was also working together with John Foster, a microbiologist at the University of South Alabama, on a microbiology textbook. The first edition of Slonczewski and Foster's *Microbiology: An Evolving Science* appeared in 2009. By 2017, it had gone through four editions. Widely assigned for undergraduate microbiology courses around the world, Slonczewski and Foster's textbook has introduced a generation of students to the field. Slonczewski herself describes it as "the field's leading book for science majors" (Slonczewski and Walton 2013, 23–24). Its pedagogical approach stresses the virtues of scientific citizenship along with discovery, the affordances of science fiction along with science fact.

In this regard, there are significant continuities between *The Highest Frontier*—indeed, all of Slonczewski's science fiction—and *Microbiology: An Evolving Science*. Like *The Highest Frontier*, the textbook emphasizes the role of speculation in shaping our modern world, galvanizing science and innovation. It also features science fiction as a resource for challenging the firmative tendencies of high-tech culture. Specifically, Slonczewski and Foster present science fiction as orienting scientific representation toward the outside and the contrariwise: the specular (i.e., the image-reflecting function of fact making and knowledge production) enmeshed with the speculative (i.e., the image-casting function of fictive extrapolation). For example:

> The resolution of the human retina (that is, the length of the smallest object most human eyes can see) is about 150 μm, or one-seventh of a millimeter. In the retinas of eagles, photoreceptors are more closely packed, so an eagle can resolve objects eight times as small or eight times as far away as a human can; hence, the phrase "eagle-eyed" means sharp-sighted. On the other hand, insect compound eyes have one-hundredth the resolution of human eyes because their receptor cells are farther apart than ours. If a science-fictional giant ameba had eyes with a retinal resolution of 2 meters (m), it would perceive humans as we do microbes. (40)

The figure of the giant ameba with its strange specular apparatus bodies forth science fiction as an epistemic practice, a way of seeing. Inviting us to consider ourselves from the outré perspective of this outlandish ameba, Slonczewski and Foster provide a heuristic for thinking about scalar comparison and the limits of visual resolution, while also encouraging students to reconsider the perspectival biases that shape human attitudes about microbes and other nonhuman entities. This little "what if?" scenario draws attention to the specificities of human perception, the situatedness of our knowledge, by compelling us to envisage the position of the alien, the monster. It affords critical observance of our world and ourselves from the vantage of science-fictional alterity.[6]

Slonczewski and Foster scatter references to science fiction throughout the textbook. The specific allusions and citations have remained fairly consistent over the different editions, despite extensive updates to the scientific content: Michael Crichton's *The Andromeda Strain* and its film adaptation; the films *Soylent Green* and *Gattaca*; the original *Star Trek* series; H. G. Wells's *The War of the Worlds*; Kim Stanley Robinson's *Red Mars*; and Slonczewski's own *A Door into Ocean*. Each citation is brief yet presented as if the student audience were already immanently familiar—tacitly encouraging a deeper engagement with the story in question, outside the textbook synopsis. Certainly, the majority of undergraduate students today are unlikely to know all the references in detail. But the textbook intimates that readers ought to investigate these fictions as part of their cultural and scientific edification.[7] For the students who do so—following the model of Jenny Ramos Kennedy, who supplements her scientific education with a host of other cultural materials and perspectives—certain passages in the textbook will take on a whole new significance.

For example, a sidebar discusses endosymbiosis in animals: "Microbial endosymbiosis has even inspired fictional creations, such as the 'breath-microbes' of Joan Slonczewski's novel *A Door into Ocean*, which acquire oxygen for human hosts swimming underwater" (28). Students who turn to Slonczewski's novel, however, will get far more than an illustration of evolutionary processes. After all, the endosymbiotic relations in *A Door into Ocean* represent a radically nonhierarchical approach to living in common, as well as an ecofeminist model for scientific practice that is communal rather than elitist, embracing a cooperative rather than dominant approach to experimentation (see Donawerth 1997; Otto 2012; Vint 2010; and Tidwell in this volume). Therefore, while the sidebar highlights a scientific fact that has inspired fiction, it actually points to a fictional relation that has—at least, in the novel—inspired an alternate science, a completely different lifeworld.

Slonczewski and Foster utilize this rhetorical technique over and again, casually mentioning a work of science fiction that allegedly illustrates a scientific principle or a fact of nature, but whose narrative shows the extent to which even purported truths might reflect habits of mind, blind spots of culture, and prejudices of humanism:

> In the early twentieth century, no one would have believed that living organisms could live in concentrated sulfuric acid, much less produce it. The *Star Trek* science fiction episode "The Devil in the Dark" portrayed an imaginary alien creature, the Horta, that produced sulfuric acid in order to eat its way through solid rock ... In actuality, of course, no creatures beyond the size of a microbe are yet known to grow in sulfuric acid. But archaea such as *Sulfolobus* species oxidize hydrogen sulfide to sulfuric acid and grow at pH 2, often in hot springs at near-boiling temperatures ... *Sulfolobus* makes irregular Horta-shaped cells without even a cell wall to maintain shape; how it protects its cytoplasm from disintegration is not yet understood. (530)

With a nod to the famous opening line of H. G. Wells's *The War of the Worlds*, Slonczewski and Foster note that conventional belief often polices the boundaries of scientific realism ("In the early twentieth century, no one would have believed").[8] The authors therefore commend *Star Trek* for daring to go where no one had gone before when it conjured the acid-spewing Horta. Reframing our current scientific understanding of archaea such as *Sulfolobus* in the image of science fiction ("*Sulfolobus* makes irregular Horta-shaped cells"), they indicate the specular limits of our knowledge ("not yet understood") but allow the actual to unfurl in the direction of the virtual, the dimension of the yet-to-come ("In actuality, of course, no creatures ... are yet known"). They perform an attentiveness to futurity and its aberrations, putting science fiction into practice at the level of language. By taking "unbelievable" things seriously—at least in the provisional form of the "not yet"—the rhetorical mode of science fiction traverses the threshold between known and unknown, soliciting further interrogation, even reconstruction. This is exactly the point of "The Devil in the Dark," after all—which again indicates how a deeper exploration of the textbook's science fiction references serves to amplify its pedagogical concerns. In "The Devil in the Dark," the crew of the *Enterprise* receives a distress call from the pergium mining colony on Janus Six. The planet is a "treasure house" of pergium and other elements, a site of intense resource speculation. As Captain Kirk says, "If mining conditions weren't so difficult, Janus Six could supply the mineral needs of a thousand planets." The standard biosensors detect no other life on the planet aside from the human colonists; nevertheless, an unknown entity has been attacking the Janus Six mining engineers in the tunnels below ground, dissolving their bodies with a powerful acid. Captain Kirk concurs with the mining engineers that this mysterious creature must be destroyed, not merely for the sake of the mining colony, but for the prosperity of the interstellar Federation. Espousing the logic of firmative speculation, the value prospects of maintaining the operations on Janus Six, Kirk says to the Chief Engineer: "A dozen planets depend on you for pergium for their reactors. They're already screaming." In this regard, the mysterious creature is nothing but a devil that disrupts the energy market and its projections of profitability. Kirk insists there can be no sympathy for the monster: "Mister Spock, our mission is to protect this colony, to get the pergium moving again."

But Spock has other ideas. Reluctant to commit to a position without empirical evidence ("I'd prefer to cogitate the possibilities for a time"), Spock nevertheless hazards alternatives to the known and the familiar: "Life as we know it is universally based on some combination of carbon compounds, but what if life exists based on another element? For instance, silicon." Kirk encourages him to press further, heedless of empiricism: "Speculate!" Kirk also begins to question his own assumptions, especially once confronted with indications of the Horta's intelligence. It turns out that the Horta is not a vicious killing machine but instead a frightened mother defending her eggs from the mining engineers, who have destroyed thousands of eggs without realizing it. When Spock performs a Vulcan mind meld with the Horta, he channels its thoughts: "Murder. Of thousands. Devils! .... Stop them. Kill! Strike back! Monsters!" So who is the real "murdering monster," the "devil in the dark"? Clearly, it is a matter of perspective. Unable to recognize the Horta's mineral existence as a viable form of life, oblivious to the silicon spheres scattered throughout the caverns as embryonic Hortas, the mining engineers literally could not see what was right before their eyes. Kirk says to the Chief Engineer, "Those round silicon nodules that you've been collecting and destroying? They're her eggs." The engineer responds, "We didn't know. How could we?" The specular horizon of knowledge, recognizing only carbon-based life (as McCoy puts it, "Silicon-based life is physiologically impossible") and insisting that no organism could tunnel through rock by excreting acid, had foreclosed the possibility of indigenous life on Janus Six in advance.

Once unshackled from their cognitive blinders, however, the colonial engineers can now see a future with the Horta, forging an alliance that allows both species to thrive. As Kirk says: "She and her children can do all the tunneling they want. Our people will remove the minerals, and each side will leave the other alone." This is all quite agreeable to the

Horta—there are plenty of minerals on Janus Six to go around, enough to sustain a thousand other planets—and when her eggs hatch, the young Hortas exuberantly assist the mining engineers: "First thing the little devils do is start to tunnel. We've already hit huge new pergium deposits. I'm afraid to tell you how much gold and platinum and rare earths we've uncovered." By opening to strange liaisons and unusual kinship relations, humans and nonhumans discover new ways to flourish in common, to live long and prosper (cf. Haraway 2016; Weston 2016).

"The Devil in the Dark" dramatizes themes that Slonczewski has often explored in her own fiction and in her scientific writings, including the affirmation of otherness—other organisms, other ways of being—and the capacity for unexpected alliances to evolve. Recall the anthrax elevator and the Mary ultraphyte colony of *The Highest Frontier*, the symbiotic breathmicrobes of *A Door into Ocean*, or the cerebral micros of *Brain Plague*. Similar themes pervade *Microbiology: An Evolving Science*, which repeatedly indicates how biological systems change over time, how parasitic relations might become mutualistic, and how human perspectives on biotic others might flip around by considering the "what if," the erstwhile unimaginable.

For example, Slonczewski and Foster write, "Despite their lethal potential, viruses have made surprising contributions to medical research.... Even lethal viruses such as HIV are being developed as vectors for human gene therapy" (184). The idea that HIV could become a mechanism for enhancing human life—HIV and humans persisting together, under conditions of mutualism rather than parasitism—features prominently in *The Highest Frontier* as well. Jenny must remember to take her regular HIV doses: "HIV was 'human improvement vector,' the original AIDS virus tamed to guard her health.... HIV made her genes fight cancer and cataracts. At Frontera, it would tune her cochlea for the space-hab rotation" (22). As Slonczewski suggested in 2013, the evolution of HIV as an adaptive "human improvement vector" is well underway in our own world: "*Highest Frontier* was written a while ago, but just this past year we've seen reported the first cure of a cancer patient using an HIV-derived vector ... And now, today, HIV, as I predicted, has become the number one most exciting vector for gene therapy" (qtd. in Brown 2013, 960). To be sure, in 2017, the Novartis pharmaceutical company released Kymriah, the first commercially approved gene-modified cell therapy, developed for treating acute lymphoblastic leukemia in children. Kymriah uses designer HIV to reprogram the patient's own immune system. It represents one of many emergent HIV genetic therapies, galvanizing the biotech industry with promises of a post-oncology future. The viral adversary, the focus of so much risk discourse and prophylactic anxiety, now presents a different kind of speculation altogether.

But things can evolve in perplexing ways, frustrating even the most forward-looking corporate gambits. Slonczewski has playfully reiterated this idea in her short fiction "Tuberculosis Bacteria Join UN." Published in the journal *Nature* in 2000, it poses as a scientific news article from the future:

> After all, the use of pathogens such as adenovirus and HIV as recombinant vectors was ancient history. Why not build supercomputers into some of humankind's most successful pathogens?
>
> *M. tuberculosis* was a prime candidate—it inhabits the human lungs for decades, in the ideal position to seek and destroy any pulmonary cells transformed by inhaled carcinogens. Tobacco companies poured billions of dollars into developing cybernetically enhanced, cancer-sniffing TB [tuberculosis].
>
> What no one anticipated was that the enhanced bacteria, like so many macroscale robotic entities in the past century, would develop self-awareness and discover a true brotherly love of their human hosts. (1001)

Letting preconceived notions go in favor of better ones, allowing for things that "no one anticipated" despite the best efforts of speculation, keeping open to the unimaginable, the anomalous, including the possibility that the enemy alien might become ally and friend—these lessons propagate across the pages of *Microbiology: An Evolving Science*, revisiting the abiding concerns of Slonczewski's fiction in the context of scientific pedagogy.

Responsible Speculation

Microbiology: An Evolving Science repeatedly endorses speculation as an essential element of the scientific career, while also attending to its limits. For Slonczewski and Foster, thinking outside the boundaries of consensus knowledge, venturing to address potential applications and implications ahead of time—even at the risk of misdirection—unbinds the shape of things to come. For instance, they discuss the discovery of rotary biomolecules:

> The idea that a living organism could contain rotating parts was highly controversial when such parts were first discovered in bacterial flagella ... Before this discovery, scientists had believed that living body parts could not rotate and that only humans had "invented the wheel." The discovery of rotary biomolecules has inspired advances in nanotechnology, the engineering of microscopic devices. For example, a "biomolecular motor" was devised using an ATP synthase F1 complex to drive a metal submicroscopic propeller ... In the future, such biomolecular design may be used to create microscopic robots that enter the bloodstream to perform microsurgery. (85)

While a failure of scientific imagination may have obscured the prospects, indeed, the very existence of rotary molecular motors for so long, Slonczewski and Foster indicate instead how extrapolative thinking has since opened up the field of nanotechnology, encouraging researchers to repurpose biological machines for engineering applications once considered impossible. Envisaging a future in which intravenous nanobots carry out precise medical operations inside the human body, they recuperate an old trope of science fiction—the so-called *Fantastic Voyage* scenario—which not only prefigured but in some ways contributed to the emergence of nanotechnology as a research field (Milburn 2015; Horton 2013; York 2015). In this way, they take a stand in ongoing debates among nanoscientists about whether such nanobots are technically feasible or simply the stuff of wild fantasy—a debate that is ultimately about the epistemological function of science fiction as such, which is to say, the science-generating capacity of speculative narratives.

Indeed, Slonczewski and Foster emphasize that adventurous, even outlandish hypotheses might be fair game, especially for addressing some of the biggest puzzles in science, such as the origin of life. When it comes to speculating, after all, we can perhaps never go far enough to overcome the limits of the speculative imagination itself. In discussing the microbial history of life on Earth, for example, they introduce the possibility that perhaps life did not start on this planet at all:

> The idea that life-forms originated elsewhere and "seeded" life on Earth is called **panspermia**. Theories of panspermia remain highly speculative. One hypothesis is that microbial life originated on Mars and was then carried to Earth on meteorites.... A Martian origin, however, only gains us about half a billion years; it does not really explain the origin of life's complexity

and diversity. Did life-forms come from still farther away, perhaps borne on interstellar dust from some other solar system? (646)

This form of open-ended questioning recurs frequently in the textbook, modeling for students how unclenching the imagination can facilitate experimental programs and exploratory research. In considering whether life could exist on the Jovian moon of Europa, Slonczewski and Foster recall that, due to Europa's distance from the sun and the thick crust of ice covering its surface, this possibility would require some mechanism of supporting life without photosynthesis. Various hypothetical schemes have been imagined by astrobiologists and science fiction writers: "These schemes are highly speculative—but tantalizing enough to encourage further NASA missions to take a closer look at Europa and its sister moons" (857). Under the right circumstances, in other words, flights of fancy may lead to flights of science.

To help students exercise their imaginative capacities and observe their own limits, each chapter of the textbook features a set of provocative questions, several of which encourage speculation. In some instances, the purpose is to puzzle out likely solutions based on known information, to apply the knowledge already in hand. For example: "Speculate why rubisco catalyzes a competing reaction with oxygen" (556). Yet in other instances, the purpose is to go beyond what is known or fully knowable in the present. When introducing the prospect of terraforming or geoengineering for redressing the problems of global climate change, Slonczewski and Foster write, "The role of iron limitation has led to speculation that we could remove excess CO_2 from the atmosphere by fertilizing the ocean with iron, accelerating growth of marine phytoplankton.... Do you think that iron fertilization would succeed in maintaining lowered atmospheric CO_2? What consequences might you predict for widespread iron fertilization?" (852). The textbook invites students to consider such questions, precisely to acknowledge the significance of uncertainty and the possible long-term implications of choices we make in the here and now. The speculative orientation of the book, then, while concerned with the technical effects or innovation pathways triggered by contemporary research, is equally about the entrainment of scientific ethics, practices of responsible research and innovation, and anticipatory engagement with the challenges of our high-tech future.[9]

Thus, science fiction takes a privileged role in the textbook as a primer for responsible speculation. For instance, in the opening pages of the first and second editions of the book, Slonczewski and Foster contextualize the history of microbiology research with a discussion of *The Andromeda Strain* and its film adaptation:

> In the twentieth century, the science of microbiology exploded with discoveries, creating entire new fields such as genetic engineering. The promise—and pitfalls—were dramatized by Michael Crichton's best-selling science fiction novel and film, *The Andromeda Strain* (1969; filmed in 1971). In *The Andromeda Strain*, scientists at a top-secret laboratory race to identify a deadly pathogen from outer space—or perhaps from a biowarfare lab ... The film prophetically depicts the computerization of medical research, as well as the emergence of pathogens, such as the human immunodeficiency virus (HIV), that can yet defeat the efforts of advanced science.
>
> Today, we discover surprising new kinds of microbes deep underground and in places previously thought uninhabitable, such as the hot springs of Yellowstone National Park ... These microbes shape our biosphere and provide new tools that impact human society. For example, the use of heat-resistant bacterial DNA polymerase (a DNA-replicating enzyme) in a technique called the polymerase chain reaction (PCR) allows us to detect minute amounts of DNA in traces of blood or fossil bone. Microbial technologies led us from the discovery of the double helix to the sequence of the human genome, the total genetic information that defines our species. (6)

This excursus on *The Andromeda Strain* rehearses the notion that science fiction anticipates and forecasts ("prophetically depicts") future developments. Slonczewski and Foster also indicate how the genre gives scientists resources for thinking askew. After all, to the extent that *The Andromeda Strain* indicates the existence of microbes "in places previously thought uninhabitable," it encourages microbiologists to search for life in extreme environments, discovering new organisms that might provide molecular tools for biological science, as exemplified by the history of PCR. At the same time, *The Andromeda Strain* suggests the need for a precautionary approach to terrestrial extremophiles as much as extraterrestrial organisms: "Extremophiles may provide insight into the workings of extraterrestrial microbes we may one day encounter ... Our experiences with extremophiles should alert us to the dangers of underestimating the precautions necessary in handling extraterrestrial samples. For example, we should not assume that irradiation would be sufficient to sterilize samples from future planetary or interstellar missions. Such treatments do not even kill the extremophile *Deinococcus radiodurans* found on Earth" (150). *The Andromeda Strain* therefore dramatizes a serious risk scenario. Whether taken literally or allegorically, the alien outbreak indicates the threat of all diseases "that can yet defeat the efforts of advanced science," presenting a template for real-life research practices: "While the details of the pathogen are imaginary, the film's approach to identifying the mystery organism captures the spirit of actual investigations of emerging diseases" (7, Fig. 1.3).

Yet even while prognosticating on innovations to come, depicting the risks of emergent diseases, and provoking scientists to think outside the prevailing limits of knowledge, *The Andromeda Strain* also stages the essential tension between firmative and affirmative modes of speculation. Indeed, Slonczewski and Foster indicate this tension in their summary of the plot: "scientists at a top-secret laboratory race to identify a deadly pathogen from outer space—or perhaps from a biowarfare lab." In the film, the true purpose of the top-secret Wildfire laboratory is unknown—perhaps even to those who designed it. In any case, for the scientists desperately trying to understand the nature of this organism, there can be only speculation. On the one hand, they look outward to otherness, to things we do not yet understand, beyond the frontier of human knowledge (as Dr. Dutton speculates, "Also it could be an organism from another planet released deliberately.... To make friendly contact. A kind of messenger to show us life exists elsewhere in the universe. It could be benign in its own environment.... We can't ignore any possibility"). On the other hand, they look inward to the paranoid fantasies of the security state, the preemptive imagination of the military-industrial complex that, by anticipating the risk of a bioweapon in the hands on an enemy, might have accidentally unleashed the very catastrophe it had feared (as Dr. Dutton speculates again, "The purpose of Scoop was to find new biological weapons in outer space, then use Wildfire to develop them.... Enemy? We did it to ourselves"). The answer is never given. Instead, the story follows the scientists as they carry out their experiments, working under conditions of endless speculation, navigating between the poles of the firmative and the affirmative—and dramatizing the very real sense in which scientists may be enmeshed in institutions and practices that strive to lock down the future, overextending the present, even as their experimental work may contribute to the exploration of difference.

Each of the science fiction stories referenced in *Microbiology: An Evolving Science* reiterates this dilemma, the sense that firmative speculation and affirmative speculation are two sides of the same coin. As the uncertain commons (2013) reminds us, "this is not simply a matter of good and bad speculation." After all, these speculative modes share the same epistemological structure, representing differing approaches to the potentialities and possibilities of change. One approach "renders latent possibilities as calculable outcomes ... turning open-ended futures into more of the same; it firms the status quo in the name of change." The other approach instead "embraces uncertainty and, in so doing, remains responsive to difference, to unanticipated contingencies. Responsive to change, it takes responsibility for the future" (ch. 1). Returning to their discussion of PCR and its reliance on Taq polymerase—an enzyme of the extremophile *Thermus aquaticus*, a denizen of places "previously thought uninhabitable"—Slonczewski and Foster write:

> PCR has profoundly impacted human society. By making it possible to amplify the tiniest amounts of DNA contaminating a crime scene, PCR has changed our judicial system by providing conclusive evidence in court cases where no evidence would have existed previously. And there may come a day when individual human genomes are sequenced as a standard medical test—with profound ethical and societal implications. The hopes and fears raised by the invention of PCR-based genomic sequencing inspired the science fiction film *GATTACA*, directed by Andrew Niccol, depicting an imaginary future in which everyone's destiny is determined by his or her DNA sequence. This knowledge could impact which jobs are available to someone, whether insurance coverage can be withheld, and even whether two individuals can marry. (250)

As scientific research opens up new discoveries and innovations, even technologies that may not yet exist ("there may come a day when ...") threaten to become instruments of predatory finance and repressive governance. *Gattaca* serves here to illustrate how PCR and genomic sequencing technologies—which have not only revolutionized biomedicine but have also improved the justice system—might just as easily facilitate forms of speculation that, by forecasting genetic risks and promises, turn probabilistic data into profit-making schemes or mechanisms of social control. The textbook asks students to consider the "profound ethical and societal implications" that might arise from contemporary research, offering science fiction as a model for thinking about the otherwise obscure connections between bench-top experiments and societal circumstances. By juxtaposing the future-opening potential of biotechnology with *Gattaca*'s cautionary narrative of firmative genomic speculation, Slonczewski and Foster reinforce an ethos of scientific responsibility, speculating beyond the confines of the laboratory to consider who might benefit, who might profit, and who might suffer.

The textbook presents a tremendous number of instances where microbiology research has improved the conditions of life for people around the world. But Slonczewski and Foster take pains to indicate how that same research has also sometimes served political and economic interests, tied to particular histories. For example:

> In the food industry, *Spirulina* is classified as a form of single-celled protein, a term for edible microbes of high food value. Other kinds of single-celled protein include eukaryotes such as yeast and algae. The twentieth century saw the development of single-celled protein as a food source for impoverished populations. The yeast *Saccharomyces cerevesiae* was grown for protein by Germany during World War I, using inexpensive molasses for culture; and during World War II, *Candida albicans* was grown on paper mill wastes. Single-celled protein was later promoted by Western countries as a food for rapidly expanding populations of developing countries. This idea inspired the science fiction film *Soylent Green* (1973), in which people of a future overpopulated Earth are forced to eat "Soylent," a food supposedly based on soybeans and single-celled protein though its true source was recycled humans. (592)

Soylent Green is cited here not because it represents the successful application of microbiological science to solve social problems, but because it addresses the profit motives that have often underwritten such projects in modern history. Slonczewski and Foster recollect the infamous twist of the film, the climactic moment when Detective Frank Thorn shouts with both horror and resignation, "Soylent Green is people!" In the film, Soylent Green is advertised as "the miracle food of high-energy plankton gathered from the oceans of the world." But after examining documents recovered from the home of a murdered executive of the Soylent Corporation, Thorn's friend Sol Roth discovers that the company's own environmental research (*Soylent Oceanographic Survey Report, 2015 to 2019*) had long ago predicted the total decimation of ocean plankton—attributed to the "greenhouse effect" ("everything is burning up"), as well the dumping of industrial waste. The vast quantities of Soylent

Green produced to feed the burgeoning human population could therefore not be made from single-celled protein. Instead, the Soylent Corporation, having foreseen the end of the plankton but wanting to maintain the status quo—that is, to secure their economic privilege by continuing to grow their market, the world population—adopted a new strategy: using dead human bodies as the protein source for Soylent Green.

This grotesque revelation allegorizes the way in which economic elites have often benefitted from forward-looking scientific solutions to crises that disproportionally affect impoverished populations as well as the nonhuman environment. As Slonczewski and Foster mention, "Single-celled protein was later promoted by Western countries as a food for rapidly expanding populations of developing countries," indexing a long history of bio-focused Western development practices that, even under the guise of helping the planet and its peoples, have often perpetuated global inequali ties rather than resolving them.[10] *Soylent Green* brings the long-term systemic processes of global capitalism and its speculative machinations to a discrete punctuation—the stark realization that, in supporting neoliberal visions of endless market growth, accepting temporary techno-fixes in lieu of fundamental changes to unsustainable systems, consumers just wind up consuming each other. Once again, the science fiction example addresses science as a social practice, cautioning that even the most high-tech remedies may inadvertently extend deeper socioeconomic problems—seizing up the future rather than unfastening its utopian possibilities.

To Practice—Speculation

Echoing the Spider-Man precept for living in a high-tech world ("With great power there must also come—great responsibility!"), Slonczewski and Foster suggest how ethical principles might be gleaned from the language of fabulation, the subjunctive conditionals and thought experiments that have been the stock in trade of superhero comics, space operas, techno-thrillers, and other amazing fantasies:

> Of course, with technological power comes a sobering responsibility, for if nature has taught us anything, it has taught us that it cannot be tampered with without cost. For instance, what would happen if an *Escherichia coli* that produced human growth hormone managed to colonize the human gut? Or, worse yet, one that was engineered to make botulism toxin.

> While these possibilities may seem unlikely, future advances in biotechnology must be guided by serious considerations of bioethics and an eye for unintended consequences. (432)

There it is, the gist of science fiction, as if ready-made for reflexive technoscience: "What would happen if …?" Envision, forecast, prognosticate. Extrapolate with abandon. Tell stories about unintended consequences. Fantasize about the collateral damage of the best of intentions. Explore scenarios, draw contingencies, sketch out lines of flight. Even "unlikely" possibilities may prove all too prescient, in hindsight—implying that, when it comes to "future advances," deciding what *should* be done may be far more important than imagining what *could* be done.[11]

When describing the possibility of life on Mars, they ask: "If no life exists—or if it exists only in the form of microbes deep underground—should we consider human intervention, or terraforming, to make Mars habitable for life from Earth?" It is a question not merely of technical ingenuity, but more specifically of ethics ("should we consider"). What responsibility ought we have to a lifeless planet? What responsibility ought we have to a planet whose only native life-forms may be bacteria? Slonczewski and Foster remind us that, as with so many of the issues explored in the textbook, science fiction provides resources for examining such questions, even as technical proposals emerge: "Scenarios for terraforming, long explored in science fiction, are now receiving serious thought among space scientists." They summarize a range of normative positions and imaginary debates already staged in literature, for example, in Kim Stanley Robinson's Mars trilogy, Jack Williamson's *Terraforming Earth*, Slonczewski's own *The Children Star*, and many other texts (cf. Pak 2016). The importance of such imaginative exercises goes far beyond pragmatic concerns or hang-ups about plausibility:

> In favor of terraforming, it is argued that Mars offers enormous natural resources of potential benefit for humanity, especially as terrestrial resources are used up. Human settlements on Mars would be a major step forward for space exploration. On the other hand, it is argued that the planet Mars is a natural monument, a place with its own right to exist as such. It should be allowed to remain in its natural state for future generations to appreciate. As a practical matter, terraforming remains unfeasible for the near future.[12]

The future, of course, tends to arrive sooner or later. But this sketch of the ethical dimensions of terraforming, the question of "should we consider," is not limited to terraforming as such—or to any particular technoscientific innovation, revolution, or crisis that may yet develop. The point is that, in advance of it becoming "practical," while it still remains "unfeasible"—that is, while it is *nothing otherwise than science fiction*—deliberating on the ethical aspects of any scientific venture is crucial. The difficulty is to keep the future open long enough to make responsible choices, before it is too late.

Slonczewski has suggested that her novels challenge the complacency of presentism, the inertial force of the status quo, encouraging us to contemplate meaningful change and what she describes as "a way out." She has taken the same approach in her scientific pedagogy, employing science fiction as a resource for critical thought and responsible innovation, advising her students of the virtues of practicing science fiction even as they practice cutting-edge biology. Among her courses at Kenyon College, for example, Slonczewski teaches "Biology in Science Fiction." In 2012, the course website featured this gnomic overture: "Learn biology through science fiction. Use quantitative reasoning. Save Hometree" (Slonczewski 2012). Alluding to James Cameron's film *Avatar* and the figure of Hometree that represents the endangered ecosystems of the world, the course solicitation gathers together the core elements of Slonczewski's praxis: fiction, science, and ethics. Significantly, it does not explicate or delimit the specific relations among these three propositions, the three objectives of the course. Rather, the juxtaposition galvanizes inquiry, imaginative questioning. Exploring the potential connections is precisely the adventure of education. As for putting them in action, reaching for the highest frontier … of course, we must speculate.

With nothing less than the future at stake.

Notes

1. See Sunder Rajan (2006), Fortun (2008), Waldby and Mitchell (2006), Cooper (2008), Mirowski (2011), and Milburn (2008).
2. See Vint (2015) and Bahng (2017).
3. On the self-analyzing tendencies of science-fictional discourse, see Gunn and Candelaria (2005), Lewis (1990), and Milburn (2014). As Landon (1995) has written, "science fiction is perhaps the most recursive and most self-reflexive of all major literary movements" (xvii–xviii). On the means by which science fiction narratives observe the conditions of their own

mediation and the formal constraints of their own speculative visions, see Clarke (2008).

4. On the extent to which science fiction has both propagated and scrutinized the politics of the American frontier imaginary, see Westfahl (2000) and Kilgore (2003).
5. As Haraway (1991) has taught us, "The main trouble with cyborgs, of course, is that they are the illegitimate offspring of militarism and patriarchal capitalism, not to mention state socialism. But illegitimate offspring are often exceedingly unfaithful to their origins" (151).
6. Perhaps it goes without saying that this passage and others in *Microbiology: An Evolving Science* instantiate what Suvin (1979) famously called *cognitive estrangement*. On the cognitive dimensions of science fiction, see Suvin (1979), Freedman (2000), Parrindar (2001), and Chu (2011). On science fiction as a way of seeing and thinking in modern culture, an orientation to science and innovation, see Landon (1995), Csicsery-Ronay (2008), Milburn (2008), and Vint (2014). On the cognitive estrangement afforded by microbiology research itself, see Slonczewski (2014, 7): "Microbiology today, as a field of discovery, is more amazing than most science fiction."
7. While most of the textbook's science fiction references have persisted across its different editions, there have been a few changes. The 2013 third edition removed one of two references to *The Andromeda Strain* that had appeared in earlier editions. The remaining *Andromeda Strain* reference in this edition does not feature any plot synopsis, however, effectively making an even stronger presumption of the reader's familiarity with the story. The 2017 fourth edition removed all references to *The Andromeda Strain* but added a new reference to Michael Crichton's *Jurassic Park*. The fourth edition also removed a discussion of *A Door into Ocean* but added an explicit citation of *Red Mars*, replacing a more covert allusion to Kim Stanley Robinson's novel that had appeared in earlier editions. As a very informal assessment of the pedagogical appeal of such references, I assigned the first edition of the textbook in my 2016 "Writing Science" class (ENL/STS 164) at the University of California, Davis. I polled the 65 undergraduates in the class about their familiarity with the cited science fiction. None of the students knew the stories directly, though all were aware of the original *Star Trek* series and a few had previously "heard of" *The Andromeda Strain* and *Soylent Green*. However, all students reported that they were "highly likely" to seek out these works of science fiction now that their attention had been drawn.
8. Compare Wells (1898): "No one would have believed in the last years of the nineteenth century that this world was being watched keenly and closely by intelligences greater than man's and yet as mortal as his own;

that as men busied themselves about their various concerns they were scrutinised and studied, perhaps almost as narrowly as a man with a microscope might scrutinise the transient creatures that swarm and multiply in a drop of water" (1). Wells's description of alien intelligences watching humans from afar, as if under a microscope, likewise resonates with Slonczewski and Foster's figure of the giant ameba observing humans from its alien perspective.

9. On the anticipatory governance of technoscience and responsible innovation, see Zülsdorf et al. (2011) and Owen et al. (2013). On science fiction for scientific pedagogy and ethical deliberation, see Berne and Schummer (2005), Burton et al. (2018), and Hall (2015).
10. Notably, the source text for *Soylent Green*—Harry Harrison's 1966 novel *Make Room! Make Room!*—does not feature the cannibalistic twist, though it is equally critical of unchecked consumerism and the capitalistic roots of climate change. On the history of single cell protein in the context of the Green Revolution and Western development models, see Bud (1993); Gibson (2012). On ways in which such development models have tended to diminish food sovereignty, increase economic disparities, and impact the environment even while addressing problems of hunger and malnutrition, see Buttel et al. (1985) and Daño (2014).
11. Spider-Man first learned the lesson about power and responsibility in Lee and Ditko (1962, 11). The technocultural principle of emphasizing "what *should* be done" over "what *could* be done" comes from Neil Stephenson's 1995 novel *The Diamond Age*.
12. Slonczewski and Foster (2009, 856). While this passage in the textbook's first edition implicitly alludes to Robinson (1992), the 2017 fourth edition states the connection directly.

References

The Andromeda Strain. 1971. Directed by Robert Wise. Screenplay by Nelson Gidding, based on the novel by Michael Crichton. Universal Pictures.

Bahng, Aimee. 2017. *Migrant Futures: Decolonizing Speculation in Financial Times*. Durham: Duke University Press.

Berne, Rosalyn W., and Joachim Schummer. 2005. Teaching Societal and Ethical Implications of Nanotechnology to Engineering Students through Science Fiction. *Bulletin of Science, Technology & Society* 25: 459–68.

Brown, Paige. 2013. A Mobius Strip of Scientific Imagination: How Deep Does the Relationship between Science Fiction and Science Go? *EMBO Reports* 14 (11): 959–63.

Bud, Robert. 1993. *The Uses of Life: A History of Biotechnology*. Cambridge: Cambridge University Press.

Burton, Emanuelle, Judy Goldsmith, and Nicholas Mattei. 2018. How to Teach Computer Ethics through Science Fiction. *Communications of the ACM* 61 (8): 54–64.

Buttel, Frederick H., Martin Kenney, and Jack Kloppenburg. 1985. From Green Revolution to Biorevolution: Some Observations on the Changing Technological Bases of Economic Transformation in the Third World. *Economic Development and Cultural Change* 34: 31–55.

Chu, Seo-Young. 2011. *Do Metaphors Dream of Literal Sheep? A Science-Fictional Theory of Representation*. Cambridge, MA: Harvard University Press.

Clarke, Bruce. 2008. *Posthuman Metamorphosis: Narrative and Systems*. New York: Fordham University Press.

Cooper, Melinda. 2008. *Life as Surplus: Biotechnology and Capitalism in the Neoliberal Era*. Seattle: University of Washington Press.

Csicsery-Ronay, Istvan, Jr. 2008. *The Seven Beauties of Science Fiction*. Middletown, CT: Wesleyan University Press.

Daño, Elenita C. 2014. Biofortification: Trojan Horse of Corporate Food Control? *Development* 57: 201–09.

de Goede, Marieke. 2012. *Speculative Security: The Politics of Pursuing Terrorist Monies*. Minneapolis: University of Minnesota Press.

Delany, Samuel R. 1977. *The Jewel-Hinged Jaw: Notes on the Language of Science Fiction*, rev. ed. Middletown, CT: Wesleyan University Press, 2009.

The Devil in the Dark. 1967. *Star Trek*. Season 1, Episode 25. Directed by Joseph Pevney. Teleplay by Gene L. Coon. Desilu Productions and Norway Corporation.

Donawerth, Jane. 1997. *Frankenstein's Daughters: Women Writing Science Fiction*. Syracuse: Syracuse University Press.

Edwards, Paul. 1996. *The Closed World: Computers and the Politics of Discourse in Cold War America*. Cambridge, MA: MIT Press.

Fortun, Michael. 2008. *Promising Genomics: Iceland and DeCODE Genetics in a World of Speculation*. Berkeley: University of California Press.

Freedman, Carl. 2000. *Science Fiction and Critical Theory*. Hanover: Wesleyan University Press.

Gattaca. 1997. Directed and written by Andrew Niccol. Columbia Pictures.

Gibson, Mark. 2012. *The Feeding of Nations: Redefining Food Security for the 21st Century*. Boca Raton, FL: CRC Press.

Gunn, James E., and Matthew Candelaria (eds.). 2005. *Speculations on Speculation: Theories of Science Fiction*. Lanham, MD: Scarecrow Press.

Hall, Shane Donnelly. 2015. Learning to Imagine the Future: The Value of Affirmative Speculation in Climate Change Education. *Resilience* 2 (2). https://doi.org/10.5250/resilience.2.2.004.

Haraway, Donna J. 1991. A Cyborg Manifesto: Science, Technology, and Socialist-Feminism in the Late Twentieth Century. In *Simians, Cyborgs, and Women: The Reinvention of Nature*, 149–81. New York: Routledge.

———. 2016. *Staying with the Trouble: Making Kin in the Chthulucene*. Durham: Duke University Press.

Horton, Zach. 2013. Collapsing Scale: Nanotechnology and Geoengineering as Speculative Media. In *Shaping Emerging Technologies: Governance, Innovation, Discourse*, ed. Konrad, Kornelia, Christopher Coenen, Anne Dijkstra, Colin Milburn, and Harro van Lente, 203–18. Berlin: AKA-Verlag/IOS Press.

Jones, Jeremy L.C. 2011. Making Strange Stuff Familiar: A Conversation with Joan Slonczewski. *Clarkesworld* 61: 56–60.

Kilgore, De Witt Douglas. 2003. *Astrofuturism: Science, Race, and Visions of Utopia in Space*. Philadelphia: University of Pennsylvania Press.

Landon, Brooks. 1995. *Science Fiction after 1900: From the Steam Man to the Stars*. New York and London: Routledge.

Lee, Stan, and Steve Ditko. 1962. Spider-Man! *Amazing Fantasy #15*. New York: Marvel Comics.

Lewis, Anthony R. 1990. *An Annotated Bibliography of Recursive Science Fiction*. Boston: NESFA Press. Updated online ed., 2008. https://www.nesfa.org/Recursion/.

Milburn, Colin. 2008. *Nanovision: Engineering the Future*. Durham: Duke University Press.

———. 2014. Posthumanism. In *The Oxford Companion to Science Fiction*, ed. Rob Latham, 524–36. Oxford: Oxford University Press.

———. 2015. *Mondo Nano: Fun and Games in the World of Digital Matter*. Durham: Duke University Press.

Mirowski, Philip. 2011. *Science-Mart: Privatizing American Science*. Cambridge, MA: Harvard University Press.

Otto, Eric C. 2012. *Green Speculations: Science Fiction and Transformative Environmentalism*. Columbus: Ohio State University Press.

Owen, Richard, John Bessant, and Maggy Heintz (eds.). 2013. *Responsible Innovation: Managing the Responsible Emergence of Science and Innovation in Society*. Chichester: Wiley.

Pak, Chris. 2016. *Terraforming: Ecopolitical Transformations and Environmentalism in Science Fiction*. Liverpool: Liverpool University Press.

Parisi, Luciana, and Steve Goodman. 2005. The Affect of Nanoterror. *Culture Machine* 7. http://culturemachine.tees.ac.uk/.

Parrindar, Patrick (ed.). 2001. *Learning from Other Worlds: Estrangement, Cognition and the Politics of Science Fiction and Utopia*. Liverpool: Liverpool University Press.

Robinson, Kim Stanley. 1992. *Red Mars*. New York: Spectra.

Saler, Michael. 2012. *As If: Modern Enchantment and the Literary Prehistory of Virtual Reality*. Oxford: Oxford University Press.

Slonczewski, Joan. 1998. *The Children Star*. New York: Tor.

———. 2000a. *Brain Plague*. New York: Tor.

———. 2000b. Tuberculosis Bacteria Join UN. *Nature* 405: 1001–01.

———. 2011. *The Highest Frontier*. New York: Tor.

———. 2012. BIOL 103 Biology in Science Fiction. Biology Department, Kenyon College, Fall. http://biology.kenyon.edu/slonc/bio3/bio03syl.htm.

———. 2014. Field of Discovery. *Locus* 72 (3): 6–7, 64–65.

Slonczewski, Joan, and John Watkins Foster. 2009. *Microbiology: An Evolving Science*, 1st ed. New York: Norton.

Slonczewski, Joan, and Jo Walton. 2013. *The Helix and the Hard Road*. Seattle: Aqueduct Press.

Soylent Green. 1973. Directed by Richard Fleischer. Screenplay by Stanley R. Greenberg, based on the novel *Make Room! Make Room!* by Harry Harrison. MGM.

Sunder Rajan, Kaushik. 2006. *Biocapital: The Constitution of Postgenomic Life*. Durham: Duke University Press.

Suvin, Darko. 1979. *Metamorphoses of Science Fiction: On the Poetics and History of a Literary Genre*. New Haven: Yale University Press.

uncertain commons. 2013. *Speculate This!* Durham: Duke University Press.

Vint, Sherryl. 2010. Animal Studies in the Era of Biopower. *Science Fiction Studies* 37: 444–55.

———. 2014. *Science Fiction: A Guide for the Perplexed*. London: Bloomsbury Academic.

———, ed. 2015. *The Futures Industry*. *Paradoxa* 27.

Waldby, Cathy, and Robert Mitchell. 2006. *Tissue Economies: Blood, Organs, and Cell Lines in Late Capitalism*. Durham: Duke University Press.

Wells, H.G. 1898. *The War of the Worlds*. London: William Heinemann.

Westfahl, Gary (ed.). 2000. *Space and Beyond: The Frontier Theme in Science Fiction*. Westport, CT: Greenwood Press.

Weston, Kath. 2016. *Animate Planet: Making Visceral Sense of Living in a High-Tech, Ecologically Damaged World*. Durham: Duke University Press.

York, Emily. 2015. Nano Dreams and Nanoworlds: *Fantastic Voyage* as a Fantastic Origin Story. *Configurations* 23: 263–99.

Zülsdorf, Torben B., Christopher Coenen, Ulrich Fiedeler, Arianna Ferrari, Colin Milburn, and Matthias Wienroth (eds.). 2011. *Quantum Engagements: Social Reflections of Nanoscience and Emerging Technologies*. Heidelberg: IOS Press/AKA-Verlag.

CHAPTER 8

Wisdom Is an Odd Number: Community and the Anthropocene in *The Highest Frontier*

Alexa T. Dodd

At the end of *The Highest Frontier* (2011), college student Jenny Ramos Kennedy's roommate becomes a plant. Of course, she's not really a plant, just as she wasn't really a human roommate. Mary Dyer is an ultraphyte, a species from "off-world" that's invaded Earth, wreaking havoc when they release cyanide in response to stress. As we learn in the first pages, thousands of people have died because of the strange microbe "you didn't need a microscope to see" (9). But while the US government is obsessed with fighting the "war on ultra," the novel's denouement suggests that humans may have something to learn from these adaptable aliens; at the very least, it's clear that ultraphytes are not the biggest problem facing Earth. Rather, it's human foolishness, particularly exemplified in the presidential race and the divisive two-party system in the United States. The novel trenchantly satirizes irresponsible humans who have abused technology and resources, leading to the "Death Belt" across the middle of the United States, melting icebergs, and flooding from rising sea levels.

A. T. Dodd (✉)
Department of English, Texas Tech University, Lubbock, TX, USA

© The Author(s) 2020

B. Clarke (ed.), *Posthuman Biopolitics*,
Palgrave Studies in Science and Popular Culture,
https://doi.org/10.1007/978-3-030-36486-1_8

Nonetheless, the novel's ending offers hope that humans have the potential to come together and begin to take responsibility for each other and the planet, if they can gain the necessary wisdom. The ultraphyte Mary's desire to understand wisdom is the trait humans need to imitate; her existence as a community—a "we"—symbolizes the way in which humans can discover the wisdom to adapt to their role on Earth and beyond.

This reading will explore the novel's emphasis on community and wisdom as an answer to the dilemmas facing the future of the Earth. *The Highest Frontier* participates in the discussions about the Anthropocene as a geological and a sociological concept. The diversity of humanity complicates this idea of a community as a unified human "we" who can take responsibility for Earth. The term "Anthropocene" gives rise to a potentially dangerous, and certainly limiting, "homogenization in a we" (Yusoff 2016, 7). I will argue for the ways in which the novel opposes such homogenization, even as it advocates for community. Mary Dyer, the ultraphyte, represents both what humans have done to the Earth in their lack of conscientiousness and what they are capable of doing for it if they pursue wisdom. *The Highest Frontier*, then, relates to the planetary scientist David Grinspoon's (2016) hope for the future of Earth: a world in which humans learn to identify with the biosphere and to use technology conscientiously, and a world, in other words, where humans gain the wisdom to act collectively, as a planetary community. Finally, Slonczewski's novel is optimistic that such a future is possible only if people assume their individual responsibilities within that community, for every person, shouldering that responsibility entails a deeper understanding of what it means to be human. Ultimately, the novel suggests that solutions to Anthropocenic problems entail moving beyond self-interest, and, perhaps, beyond our planet.

The Anthropocene

As a vision of the future of Earth, *The Highest Frontier* participates in contemporary conversations about humanity's impact on the Earth, particularly in concerns about climate change, resource use, and endangered and invasive species. The spacehab of Frontera College—with its controlled climate, engineered miniature animals, and endless source of amyloid resource substitutes—offers a technological oasis from the problems facing Earth. But as Jenny's freshman year unfolds, we watch as the spacehab, too, becomes prey to these issues, from mosquitos to ultraphytes to flooding. Geologists Paul Crutzen and Eugene Stoermer coined the term Anthropocene in 2000 to suggest the arrival of a new geological epoch. The term describes "human dominance of biological, chemical and geological processes on Earth" (Crutzen and Schwagerl 2011, 1), and proponents argue that it designates humanity's irrevocable influence on the Earth. In particular, the Anthropocene coincides with and so may designate the period of increased carbon dioxide emissions, ozone depletion, rising sea levels, deforestation, and mass extinctions of species since the evolution of the human race (Alcaraz et al. 2016, 321). Crutzen (2006) points out, "In a few generations mankind is exhausting the fossil fuels that were generated over several hundred million years, resulting in large emissions of air pollutants" (14). Scientists fault the rise in "greenhouse gases"—particularly carbon dioxide, nitrous oxide, and methane—with increases in global temperature and sea levels (15). Coal burning has also led to the acidification of precipitation and of lakes, resulting in fish death and forest damage. Because of these and other issues caused by human action, many scientists agree that it is "more than appropriate to emphasize the central role of mankind in geology and ecology by using the term 'Anthropocene' for the current geological epoch" (16). Biermann (2012) states that "Humanity itself has become a powerful agent of earth system evolution" (4).

But while the Anthropocene includes matters of climate change and problems of resource use, it is not a purely negative concept and refers more broadly to the physical geological layer created by human influence. Thus, the Anthropocene, as a geological epoch, would replace the Holocene, the epoch previously used to describe the 12,000 or so years since the last ice age. Some argue that the Anthropocene started as early as the 1600s, during the Colombian Exchange, when trade between the Americas and Europe resulted in the interchange of species across the Atlantic, causing a sudden drop in carbon dioxide levels because of the "massive biological disruption" between hemispheres (Grinspoon 2016, 219). Others suggest the Industrial Revolution, with the invention of the steam engine, or the Cold War, with its above-ground nuclear tests, as starting points. Crutzen (2006) in particular advocates for the latter part of the eighteenth century because "during the past two centuries, the global effects of human activities have become clearly noticeable" (16). While proponents may differ on the exact starting date, they agree that the beginnings of the Anthropocene involved a key shift in human technology, resulting in our deeper impact on Earth's processes. The various proposed starting dates "describe a series of interesting waypoints in the development of the changing and increasing human influence" (Grinspoon 2016, 221).

Therefore, as it denotes human dominance of Earth processes, it also implies human responsibility—the need for "us" to become aware of our influence and reduce its negative effects through the conscientious use of resources and technology. Many scientists argue that "we must learn to grow in different ways than with our current hyper-consumption" (Crutzen and Schwagerl 2011, 2). Proponents advocate for more innovative energy solutions, better recycling habits, restoration of the natural environments we've harmed, and even "control of human and domestic animal population" (Crutzen 2006, 17) to reduce resource use. Grinspoon, for example, calls us to become "planetary gardeners" (412), utilizing technology and environmental policies to counter present and conceivable future problems. For example, some scientists support geoengineering solutions: using technology to intervene in Earth's natural processes, such as pumping iron into the southern ocean to stop algae depletion (Biermann 2012, 8). Many argue that we can prevent future ice ages by adding artificial gases to the atmosphere, and that, in the future, we can develop the technology to "deflect meteorites and asteroids before they could hit Earth" (Crutzen 2006, 17). Ultimately, scientists argue that taking on our role in the Anthropocene should not only involve countering the problems we have created but also implementing measures to make life better on our planet.

Furthermore, this "life" is not just human. Indeed, the "linkages between biophysical systems and social systems have grown to the point where it is necessary to speak of human-dominated ecosystems, operating at various scales, from local to global" (Alcaraz et al. 2016, 321). We can no longer view humans and nature as separate: "The Anthropocene becomes an invitation to re-think the place of humans on the planet or, ultimately, what can be considered 'human' or 'natural'" (323). Many thinkers view the Anthropocene as an opportunity for humans to begin seeing the planet as a living, interconnected set of systems—a biosphere with which we are deeply connected. In this sense, understanding the Anthropocene lays the foundation for thinking about the Gaia hypothesis proposed by James Lovelock and Lynn Margulis. According to this theory, we should not consider Earth as merely a planet on which we live, but rather a kind of living, evolving entity itself. As Grinspoon (2016) articulates the idea,

> Margulis and Lovelock proposed that the drama of life does not unfold on the stage of a dead Earth, but that, rather, the stage itself is animated, part of a larger living entity, Gaia, composed of the biosphere together with the 'nonliving' components that shape, respond to, and cycle through the biota on Earth. Yes, life adapts to environmental change, shaping itself through natural selection. Yet life also pushes back and changes the environment, alters the planet...So evolution is not a series of adaptations to inanimate events, but a system of feedbacks, an exchange. Life has not simply molded itself to the shifting contours of a dynamic Earth. Rather, life and Earth have shaped each other as they've coevolved. (67)

In other words, if Earth is a self-regulating system of living and nonliving components, and we are bound to that system, then we should contemplate our role in shaping Gaia. Later on, I will explore the significance of Gaia theory in relation to the Anthropocene and *The Highest Frontier* in greater detail.

While the Anthropocene is useful as a concept for understanding our role on the planet, many thinkers have commented on the potential controversy latent in the term. Yusoff (2016) contends that, in imagining a future geological layer, discussions of the Anthropocene assume a kind of "unity of the 'Anthropos'" (6), or a homogeneity among humans. "Future humanity," she asserts, "becomes an erasure of contemporary social differences and inequalities" (4). In other words, the idea of an Anthropocenic layer—however scientifically accurate it may prove to be—offers a vision of humanity that erases present-day diversity and inequality among societies, even individuals. In that sense, discussions of the Anthropocene have the potential to overlook those differences as they exist now. Thinkers like Yusoff, and Ernston and Swyngedouw (2018), argue that this neglect is apparent in the ways proponents of the term obliquely combine all of humanity in their narratives and undifferentiated plural pronouns: "we" have harmed the Earth, and it is "our" responsibility to right these wrongs. Such language overlooks stratifications of power that leave some humans more responsible than others, indeed make some populations victims rather than agents of environmental change. "In the quest for solutions to urgent collective action problems, the focus has primarily been on means rather than ends and attention has hereby been diverted away from social and cultural norms, practices and power relations that drive environmental problems in the first place" (Lovbrand et al. 2015, 212). In other words, the emphasis on finding solutions to the Anthropocene has created these falsely unifying narratives, ignoring the differences among societies. Such narratives can further formalize power relations: "The fundamental challenges to societal organization posed by the Anthropocene are, paradoxically, to be countered by many of the same institutions that have allowed the recent human conquest of the natural world" (Lovbrand et al. 2015, 214). Thus, solutions may inadvertently promote the power of institutions and governments while continuing to victimize those under their control, such as workers of the global south who could be disproportionately disadvantaged by environmental regulatory laws. Even this statement could be read as an overgeneralization, as there are certainly workers of the global south who could benefit from certain environmental regulations; where a farmer might be economically disadvantaged by laws against pesticides, his neighbors might benefit from the reduction of pesticide residue in their drinking water.

Unified depictions of humanity are therefore problematic: The "human species, as main driving force of the Anthropocene, is in itself utterly divided in wealth, health, living standards, education, and most other indicators that define wellbeing" (Biermann 2012, 6). As Grinspoon (2016) admits, "We are not a harmonious, coherent entity" (422). And as Haraway (1991) has written, "the production of a universal, totalizing theory is a major mistake that misses most of reality, probably always, but certainly now" (67). How, then, are we to approach a solution to the problems of the Anthropocene? If humanity must unite in this new age, how can "we" do so without overlooking societal differences? For the possible shape of an answer to these questions, I propose looking at the idea of community that animates *The Highest Frontier*. While the term "community" may imply yet another form of forced homogenization, I argue that this novel simultaneously advocates against such homogenization, particularly in its narrative form and its conclusion. As in the cyborg myth, the novel is "not afraid of permanently partial identities and contradictory standpoints" (Haraway 1991, 15). Finally, the novel's emphasis on wisdom and individual responsibility offers a more complete answer to the dilemmas of the Anthropocene.

Community

The novel centers on life at Frontera College, itself a small community. Indeed, the novel is in some ways about Jenny's initiation into that community, arriving as a nervous freshman and gradually making friends as she attends classes, joins the slanball team, and participates in clubs and events. Furthermore, her community quickly comes also to include the colonists of Mount Gilead, adjacent to the college and sharing the same spacehab. In her first week at Frontera, she volunteers with Homefair, a group that helps build houses for citizens on Mount Gilead, many of whom are Centrists, believing in the Biblical notion that the universe is a closed "firmament." Nonetheless, the group represents the gathering of volunteers of all beliefs, with donations from the Centrist First Firmament Church as well as the First Reconciled Church, which aligns with the opposing Unity party. In this volunteer experience, we can read Jenny's discomfort with the Centrists, whose beliefs she has been raised to oppose: "Jenny tried not to stare, horrified yet fascinated to see ordinary people who really believed the stars were pasted on the sky" (145). And yet, as Jenny interacts more with the Centrists in the hab, these characters take on more depth. Leora Smythe, for example, the late Centrist mayor's widowed wife, turns out to be more complex than she appears in her traditional "long pioneer dress with a round bonnet" (145). Similarly, Anouk's boyfriend Rafael is Centrist, but he cooperates with Jenny at the voting ballots to make sure everyone in the hab has the opportunity to vote. Thus, the novel avoids flattening or merely constricting characters into two opposing parties while it simultaneously brings them into a community. This is a community made up of rounded individuals, diverse in their opinions and beliefs, but similarly striving to make life work in the spacehab.

Furthermore, Jenny's participation in this community entails not only her interactions with these diverse individuals, but also her responsibility toward them. For example, as an EMS volunteer, Jenny sacrifices sleep and studying to attend to her fellow college students as well as to the citizens of Mount Gilead. When Jenny discovers the hab's potential to flood, she helps organize a demonstration to build awareness, which leads to the construction of lifeboats for both the human and animal members of the community. Her efforts (greatly abetted by those of her friends) help to save the community when the hab does flood near the end of the novel. The novel's image of community, in the college and Mount

Gilead, is therefore one of diversity as well as responsibility; the pursuits of knowledge, and of life itself, serve as the foundation.

In this light, community is not about sameness among members but about a common vision or goal. As the story demonstrates, there is conflict within this community; college president Dylan Chase has to mitigate a dozen problems with the faculty, students, and spacehab every day. But as the novel's ending reveals, most of the members of this community do share a goal: They want to make life work on this new frontier. For example, at the moment when the hab faces the blackout that causes the flood, many of the students and the colonists find themselves in the Mount Gilead courthouse waiting to vote in an election. Leora, the mayor, guides them to evacuate, following the flood drill. The students and the colonists march together to the tune of "Nearer, my God, to thee, nearer to thee": "The colonists took up the song, evidently one they knew well. It made a good walking song. Farmers marched out, and mothers out with children. The students didn't know the drill or the song, but they fell under the calming influence of those who did" (415).

In this brief moment, we have an image of a community, united in their purpose despite their differences. While many of the students do not share the same religious beliefs as the colonists, those differences do not matter—indeed, perhaps unite them even more—in this moment. While survival seems like an obvious common goal, it is, essentially, what proponents of the Anthropocene argue is at stake: the survival of our species and our biosphere. As Grinspoon argues, that commonality ought to unite us, ought to help us take a species-level view of ourselves. "It's fine to assert the importance of difference, of complexity within the human superorganism," he asserts, "but to deny the validity of a parallel and conjoined species level global narrative is regressive, and ultimately runs counter to the needs and aspirations of all peoples" (422). As the novel shows, it is possible to be united in a goal without undermining diversity.

In a similar way, the novel's narrative form creates a metaphor for the kind of community necessary in the Anthropocene. While it focuses primarily on Jenny's experience, and to a lesser degree on that of the college president Dylan, the plot is far from straightforward. We jump from Jenny's slanball practice to elephants in her amyloid cottage, from RNA-helix roller coasters to the "Feast of Fools" at Cockaigne Castle. At times, it seems as though the various strands are unrelated, as though the numerous, minute details of this story world will not fit together into a coherent whole. This is perhaps nowhere as obvious as with the ubiquitous Toynet, with its various windows popping up incessantly, distracting characters from the moment at hand, and driving the narrative away from the issues in the college to those on Earth as well. Nonetheless, this disjointedness seems deliberate. We watch as some of the most random details—Professor Abaynesh's two-headed snake, for example—begin to shed light on the novel's larger trajectory—such as the two presidential candidates eventually joining the same ticket. The various details come together like pieces in a jigsaw puzzle. Thus, we might read Father Clare's jigsaw puzzle as a metaphor for the narrative altogether. As Jenny notices,

> "The pieces…They don't all fit."
>
> "None of them fit. They come from hundreds of different puzzles."
>
> "But—" She picked up a piece tentatively. "There are holes showing through. And some look…"
>
> "Shoved in." Father Clare smiled broadly… "Life takes a lot of shoving." (86)

The narrative parts, much like the pieces in Father Clare's puzzle, gradually come together to form a larger picture. Some of these pieces need interpretive shoving only because the story is so richly meticulous. This seemingly disjointed but richly interconnected narrative form creates a metaphor for the kind of community humanity must become to address the problems of the Anthropocene. The narrative offers a diverse array of strands but ultimately ties them together into a unified story. Similarly, a worthwhile community allows for—indeed, needs—diversity among its members while uniting them in a goal. To put it another way, our species-level community may not fit together perfectly—it may involve some shoving (respectfully) of ideas and ethics—but hopefully we can form a common vision of our role on Earth. Similarly, though Toynet seems to sever the storyline into snippets, within the story the technology serves to connect individuals across the world and into space. Slonczewski's projection of information technology does not lament the possibility of a hypermediated world. Rather, it offers a glimpse at the potential for technology to create larger communities, where more people's thoughts can—quite literally—be seen and heard.

Wisdom and Humility

Embedded in the novel's image of community is this necessity of diversity; the text may therefore offer a solution to the "homogenization in a we" that Yusoff and other critics of the term "Anthropocene" delineate. But such diversity is not merely meant to embrace undeniable human differences in a form of what political philosopher Kymlicka (2015) identifies as neoliberal multiculturalism, which involves a kind of erasure of those differences in the name of a false equality (7). Rather, meaningful diversity involves different perspectives, as well as an awareness of those disparities so that diversity becomes the starting point for something greater, rather than an end in and of itself. In *The Highest Frontier*, this kind of diversity opens up the possibility of wisdom. Wisdom understood as a biological as well as intellectual characteristic becomes an important thread throughout the novel. Professor Abaynesh recruits Jenny's help to research the possibility of creating wise plants:

> Animals are just plants with wanderlust. Listen: Half a billion years ago the Cambrian explosion left polychephalic fossils, three-lobed creatures with five eyes and a long snake-like nose. Why do we vertebrates have it all in one head? Ever since we evolved, we pigheaded vertebrates have messed up the world, refusing to see anything outside our single heads...Ari [a plant] has a nervous system with ten or twenty flower heads. Why not us? If we had two heads—might we see two points of view? (85)

Thus, Abaynesh's search for a wisdom gene intuits the necessity of multiple perspectives. Later, Abaynesh shows Jenny how she is making a "wisdom network, nerves that can be organized to guide wise and cooperative behaviors" (168). The nerves are meant to help the plants learn to "share light and grow equally, rather than overcrowding" (168). Wisdom, therefore, also involves cooperation. It is not an accident that Abaynesh uses words like "share" and "overcrowding," vocabulary that could also aptly be used in discussions of the Anthropocene. A wise community, then, is one in which multiple perspectives are considered. As Lovbrand et al. argue, the conversation around the Anthropocene needs to focus less on solutions and more on opening up "conceptual and political space where a diversity of green diagnoses, comprehensions and problematizations can be debated and contested" (216). While searching for solutions is important, they argue, we must first have the conversations that allow for diverse opinions to help shape those solutions. Grinspoon (2016) argues a similar point: The belief that we cannot recognize valid differences while also coming together as a species to find solutions is "one of those false dichotomies, amplified by binary political thinking, that interfere with sensible approaches to our global problems" (422).

Furthermore, the novel ultimately posits that such binary political thinking—and thus, elections under the auspices of a two-party democratic system—is ineffective precisely because either party is closed off to alternate perspectives. The two parties represent a false binary in the sense that they claim to hold mutually exclusive ideas; as a result, voting threatens to become "a broken system" (413), as the candidates split the country fifty-fifty. In a sense, both parties seek a "unitary identity and so generate antagonistic dualisms without end (or until the world ends)" (Haraway 1991, 65). In reality, there are more than two ways of looking at the world; likewise, there is not one side that holds all the wisdom and another all the folly. As Jenny realizes when Mary asks what wisdom is: "It was surprisingly hard to think of anything or anyone in Frontera who never had a wise moment" (299). Despite foolish behavior, everyone has the potential to be wise; their perspectives are therefore worthwhile.

To further understand the novel's depiction of wisdom, let's look at the presidential debate in the latter part of the novel. Jenny and her friends manage to smuggle Mary's reverse control wisdom plants into the presidential debate, under whose aromatic influence both candidates begin to realize the narrow-mindedness—or at least the flaws—of their own political party. Before spouting party rhetoric, Unity candidate Anna Carillo and Centrist candidate Gar Guzman both pause to think. Between responses, they comment on what their opponent gets right, on the ways that they agree. As a result, the wisdom plant inspires the two candidates to join tickets. Gar becomes Anna's running mate because of their mutual realization of the need for a "bipartisan effort" (390). While the novel shows that the Unity party is the wiser party in terms of policy, it suggests that true wisdom should generate cooperation. Together, Gar and Anna become two heads—two perspectives—on a single body; the novel is hopeful that they will govern with wisdom, cooperating much like Abaynesh's multi-headed plants.

Furthermore, the multi-perspectives lay the foundation for much-needed humility. At the debate, Anna asserts that the Firmament theory is wrong because it is arrogant:

> Even if the 'Firmament' could be consistent with an honest understanding of the heavens—it would be wrong. It's a belief totally self-centered. How can we teach children that the entire universe revolves around our own selfish existence? When God so clearly expects us to grow beyond what we are now. (375)

Amazingly, the previously impervious Gar agrees with her, realizing their need to grow "beyond even the Firmament" (375). Let us apply this parable to the matter of governance in the era of the Anthropocene. As Grinspoon points out, much of the controversy around the Anthropocene is that it smacks of "self-centered delusions about our own significance" (207). Science has heretofore tried to lead us out of such delusions. Therefore, in order to truly take responsibility of the Earth, we must do so with humility, in the realization that though we have acquired a certain technological dominion, we must not place our own interests above that of the biosphere. Individuals gain humility only when they recognize the limits of their own viewpoints and interests and begin to consider the perspectives of both human and nonhuman others. As the geographer Hulme (2014) points out, "Humility draws attention to the limits of scientific knowledge, especially such knowledge of the climatic future with its essential openness and indeterminacy" (305). Humility is therefore a trait of wisdom. *The Highest Frontier* suggests that only through such humble wisdom can we work in harmony with each other and with the Earth in order to address the problems of the Anthropocene, so as to "reach for the stars" (432).

Mary the Ultraphyte

Furthermore, the ultraphyte Mary offers us a deeper understanding of the novel's stance on community and wisdom. As a microbial organism, a "quasispecies," Mary exists as a community, as a "we." When Jenny first meets her new roommate, Mary uses the plural personal pronoun. At first, we assume that her strange comments have to do with her alleged autism; it is only later that we realize that her quirkiness is due to being an ultraphyte. Mary misses the orientation pow-wow the day she meets Jenny, and explains, "We were delayed … The vote was close … Twenty-one to twenty-two. It took a long time" (47). At the time, Jenny assumes this is an incomprehensible "aspie" joke. Later, we learn that ultraphytes must "vote" in order to take action. Therefore, in order to avoid being paralyzed, they have to contain an odd number of cells. Abaynesh explains: "They're not simple cells, of course, more like 'citizens' of a colony. A colonial organism, in which each individual casts a 'vote.' The whole group takes a vote, a hundred times a second. So, an even number of 'cells' is bound to reach a tie vote soon. Then they're paralyzed" (197–98). In that, they must vote in order to act, ultraphyte behavior gestures toward the complications of democracy. Furthermore, ultraphytes suggest why a two-party system has the potential to paralyze a democracy; just as an ultraphyte needs an odd number of cells, a democracy needs more than two parties that, in supposedly dividing a population in half according to beliefs, thereby prevent action. For ultraphytes, wisdom is an odd number.

Mary's pursuit of wisdom—what wisdom means for humans—is the trait humans most need to imitate. At the Feast of Fools at Castle Cockaigne, Mary tries to figure out if the students' charades are wise. When Anouk performs a dance that projects virtual numbers, Mary texts Jenny, "**WISDOM?**" Jenny responds that it is both wise and entertaining (328). After another skit pokes fun at the current Centrist president, Jenny texts Mary, "**HUMOROUS AND WISE**" (329). But a few moments later, the students jokingly call for a contest between a bear and an ultraphyte: "Suddenly there were gasps. Along the floor snaked a long, sallow shape with regular eyespot cells of an ultraphyte" (330). Jenny assumes that it is an object made out of the all-purpose protein amyloid, but Professor Abaynesh orders Mary to get back to the lab. Arguably, Mary has assumed that all of the performances at the Feast are wise after Jenny's texts. So, when the students jokingly called for the appearance of an ultraphyte, Mary assumes that it would be wise of her to unmask herself as an ultraphyte. Despite her misjudgment, we here see Mary actively attempting to understand wisdom; perhaps amusingly, she has a rather difficult time learning from college students. Nonetheless, Mary pursues wisdom (as well as humor) in an attempt to understand humans—in an attempt to *be* human. As Abaynesh explains when the truth about Mary is discovered, "Mary was a community…A whole research lab of ultras, trying to be a human" (388). While Mary fails to pass as a person, her attempt offers perspective on what it means to be human. Early on, Mary says to Jenny, "Tell us how to be human" (112), indicating her objective before we know what she is. While she takes lessons on social skills from Dean Kwon, Mary most often tries to learn what it means to be human from

Jenny. Her focus on wisdom, therefore, is not just arbitrary. To have wisdom, the novel implies, is to be truly human. In the character of Mary, we therefore have the novel's explicit emphasis on the trait humans need—wisdom—in order to call ourselves human.

Furthermore, the alien being Mary's pursuit of wisdom influences her human model Jenny. When Mary asks, "For humans, what is wisdom?" (299), Jenny at first struggles to offer a definition. "'Wisdom is…when people act wise,' she finished lamely. 'When they make the best choice. In truth, wisdom is hard to find in Frontera'" (299). Mary's question spurs Jenny to think more deeply about what wisdom means. She begins to wonder if wisdom is always good. As she asks Father Clare, "[W]hat if an evil person became wise? What if, say, they wanted to blow up Earth; would they become wise enough not to do it, or would they just learn a wiser way to do it?" (367). While Father Clare admits that he doesn't truly know, his response is indicative of what wisdom means: "If destroying Earth is wisdom, I'm a fool" (368). We know—from Jenny's words and his character—that Father Clare is no fool. Therefore, destroying the Earth can't be wise. While Jenny's question is hypothetical, we might read Father Clare's answer in the context of the Anthropocene: Wisdom entails not destroying the Earth. To be wise—to be human—means taking steps to save the Earth. Further down, I will show how our humanity is the central matter at stake if we do not learn how to respond to the problems facing Earth.

Of course, because Mary is an ultraphyte—an invasive species responsible for the deaths of thousands of Americans—her presence at Frontera is complicated. According to one of the professors, Mary is a "defense experiment" (389), a biological weapon. And although that is the final explanation the text offers for Mary's existence, the story as a whole suggests she is more complex. As Jenny says, "Mary never seemed like a biological weapon. Just a mixed-up *companera*" (389). And in the end, Mary offers a solution to the spacehab's solar power sourcing. The spacehab relies on microbes in its outer shell that turn sunlight into power (16); but because the salt-eating microbes can't absorb enough light, the hab must also rely on power pumped up from Earth. Therefore, when Earth cuts off the spacehab's power near the end of the novel, Frontera floods. But when Mary earlier escapes into the outer shell, she reverts into single cells that are more efficient at absorbing sunlight. Mary's presence in the outer shell effectively saves the school after the flood and offers a future solution to its energy sourcing. As the director of Homeworld

Security explains, "Ultraphyte photochemistry could boost our solarrays, enough to convince Congress to build them" (438). Thus, though Mary, or rather, the ultraphytes, originally pose a threat to the future of Earth, they evolve to take on a symbiotic role.

Gaia and Human Responsibility

Mary, as an ultraphyte, can serve as yet another example for humans. Humans, like ultra, have harmed the Earth. But if we can learn to adjust to our new role as the dominant species and become a wise community, maybe we can save Earth. This link ties *The Highest Frontier* to the vision of a Wise Earth, what Grinspoon (2016) calls "Terra Sapiens" (276):

> This is an Earth where we and the planet have both changed to come to grips with each other; where we've learned to live comfortably over the long haul with world-changing technology, applied with a deep understanding of planetary function; where intelligent and wise application of our engineering skills has become smoothly integrated into global processes. It's a vision for the planet, but it's also an aspirational name for ourselves, for who we must become to manifest this world. On Terra Sapiens we won't differentiate between the two because we'll identify deeply with the planet. We'll understand that wise self-management and wise planetary management are one and the same. (276–77)

In this vision for the Anthropocene, humans take on responsibility for the planet with the wise use of technology and resources. Humans not only learn to take a "species view" of ourselves, recognizing our common goal to come together as a community; humans also learn to identify with Earth.

That identification relates to Lovelock and Margulis' Gaia hypothesis, the theory that Earth is a kind of living being. Much like an ultraphyte, Earth is a community of organisms, a set of feedback loops that work together to keep the planet functioning. As a name for the hypothesis, "Gaia" is, of course, a metaphor: In ancient Greek mythology, Gaia is the Earth goddess, the mother of all life. In choosing the mythological nomenclature, Lovelock and Margulis wanted to imply Gaia's self-regulating capacity. As Lovelock described it, Gaia is a "system able to homeostat the planet" (qtd. in Clarke 2017, 3). But as thinkers such as Bruno Latour have pointed out, Gaia has often been misunderstood in the scientific community, wrongly characterized as a "mother earth" projection, an almost conscious being "ruling over the earth for the benefit of all organisms" (63). More accurately, Gaia might be characterized "not as living but as metabiotic, in that it arises from the systematic coupling of living operations with the abiotic dynamics of their cosmic, solar, and Earthly elements" (Clarke 2017, 22). In other words, Gaia is a name for all of Earth's processes, both living and nonliving, as they work together in a systemic ensemble that keeps the planet functioning in a viable state.

So why not simply use the word "earth system" to describe the theory, as some scientists have suggested? As Latour explains it, Gaia is a useful term precisely because it is difficult to pin down: The "very success of the prefix 'Gaia' makes it difficult to stabilize it" (Latour 2017, 62). "Gaia" suggests a more dynamic process than "earth system," asking us to consider the Earth as a vibrant, perhaps even unpredictable, place. At the same time, Gaia should not be considered an "already constituted totality" (Latour, qtd. in Clarke 2017, 6), a unified whole that the word "system" might suggest. This suggestion is problematic because, like the unifying we of the Anthropocene, it implies a false homogeneity of Earth's processes, a kind of fixed, stable order that cannot be disrupted. Rather, the term "Gaia" indicates the heterogeneity—the diversity, even instability—of Earth's parts. The whole is not greater than the sum of those parts. In this way, while we are part of Gaia, we are also separate from it: "we belong and don't belong, are both a part of and apart from its processes" (Clarke 2017, 9). Therefore, Gaia constitutes a proper understanding of our role on Earth in the Anthropocene: We should not assume that the Earth can simply adjust and self-regulate regardless of our actions. Nonetheless, Gaia may help direct us "toward the moral of our responsibility" (Clarke 2017, 8), toward a realization that our actions can have grave consequences, as well as powerful possibilities. To put it another way, the Gaia hypothesis encourages us to have a communal mindset toward the whole Earth, even as it prevents us from forgetting our specifically human responsibilities in that community.

As I've already shown, this notion of responsibility, of taking on one's role, is also tied to the notion of community in *The Highest Frontier*. Jenny assumes responsibility for others in her community. While the novel emphasizes the importance of community, it focalizes on Jenny, an individual. As good fiction necessitates a protagonist, *The Highest Frontier* offers us a strong female lead through which to understand this world. But we can view Jenny's individuality in more nuanced terms: "In

The Highest Frontier, Jenny emerges as a leader within the context of a broader group of students and Earth community who desire change" (Slonczewski 2019). Jenny takes on her individual responsibility—one we might even say she has been genetically destined to assume, with "three presidents and four senators in her family tree" (2011, 13). As Jenny tells the news anchor Clive, her "family has a long tradition of leadership protecting Earth's precious global environment" (13). While statements like these might come from a press prompt, and while Jenny's genes might have been determined by genetic engineering, Jenny still takes her role as a leader to heart, wanting desperately to do what is right by her family and the Earth.

Moreover, Jenny's particularities—the things that make her an individual—are what make her a good leader. Her "public mutism," for example, "a gene missed in the embryo" that gives her fear of public speaking (13), actually becomes an asset because it enables her to sympathize with Mary's awkwardness: "Talking to Mary was easy, she realized. She picked up right away that this Mary was a lot more challenged than herself" (47). Where another student might have tried to ignore Mary because of her supposed "aspie" traits, Jenny accepts her as her friend and tries to help her. And, as we've seen, it is partly through Jenny's example that Mary learns wisdom and ultimately helps save the hab. Jenny's individuality, then, helps her solve the problems facing Earth and the spacehab. Thus, while the novel illustrates the kind of community needed to face the problems of the Anthropocene, it still emphasizes the importance of each individual. Indeed, a community can only make a difference if the individuals who compose it shoulder their distinct responsibilities. As the psychologist Peterson (2019) asserts, the individual is "the locus of responsibility" (20–21). In this sense, *The Highest Frontier* offers another example of how to face the real problems of the Anthropocene: Each of us must embrace our individual responsibility toward the Earth. Only then can a community, a global mindset aimed at promoting planetary well-being, begin to form. As Slonczewski succinctly articulated, "Social movements achieve change; but they need inspiration and leadership of individuals" (2019).

In this light, the novel's setting in a liberal arts college takes on deeper significance. As Dylan asserts in his welcoming speech to Frontera College's new freshmen, education "is the highest calling of the human race" (2011, 40). While Dylan's speech is certainly a satirical rendering of college president vernacular, the novel genuinely illustrates the potential of education to foster the pursuit of knowledge and, as I've shown, wisdom. Indeed, Frontera's classrooms are vivid examples of the great adventure of education—from the intellectually rigorous dialogues in Politics class to the virtual reality "quests" in Life lab. We watch as Jenny and her classmates take lessons from class into their conversations in the cafeteria and social gatherings. By the end of the novel, through her adventures in and outside of the classroom, Jenny has attained a fuller sense of self: She is finally able to say good-bye to Jordi, her late twin brother whose virtual ghost she kept visiting in ToyWorld. Thus in the act of forming a community and discovering wisdom, Jenny also becomes more comfortable in her individuality. In stepping into a role of leadership, in taking on her responsibility, Jenny discovers who she is. Through academic study, "you wake up and realize who you are, and then you're ready to take the world on your shoulders" (Peterson). Through her experiences at Frontera College, Jenny is becoming who she is meant to be. To put this self-discovery and actualization in the vocabulary of thinkers like Grinspoon and Hulme, Jenny is becoming more human.

It is precisely this enterprise, the endeavor of becoming fully human, that the Anthropocene thrusts upon us. Our very humanity is at stake if we do not learn how to address the problems we've created. "What makes us human, more than anything else, is our ability to work together to modify our environments in creative ways, powered by abstract thinking" (Grinspoon 2016, 435). Thus, in forming a global community—the very solution *The Highest Frontier* suggests is necessary for addressing the problems of the Anthropocene—we also become more human. Hulme (2014) similarly argues that discussions on caring for the planet should start with discussions on what it means to be human:

> Rather than putting science, economics, politics or the planet at the center of the story of climate change I am suggesting that we put the humanities—our self-understanding of human purpose and virtue—at the center. When we talk about climate change we should not start with the latest predictions from the climate models, nor whether we have passed some catastrophic tipping point.... We should start by thinking about what it means to be human.... What I am suggesting here is that we need a more explicit Aristotelian contemplation on the good life, the nature of well-being and the cultivation of virtue. The question then becomes less "the world we want" than it is "the people we should be." (308–9)

We should pursue the virtues of wisdom, humility, faith, hope, and love as a way of responding to Anthropocentric challenges. Cultivating these classical and Christian virtues "is something that all of us humans can do. Indeed, perhaps it is something that all of us are called to do...to sustain the mutual evocation of goodness and human dignity that alone transcends our diversity. Climate change provides us this opportunity" (Hulme 2014, 309). While I have not addressed faith, hope, and love in *The Highest Frontier*, these virtues are also woven in throughout the novel. It is no accident that Father Clare, a Christian priest, is the novel's moral center. On a Sunday morning at All Saints Church, Jenny listens to Father Clare's sermon on love and hope: "Remember, God's answer to the problem of evil is compassion...Our task is to love God and love our neighbor. Through our love, the barriers that divide us will crumble; our divisions being healed, we may live in justice and peace...Above all, hope. 'For in hope are we saved'" (154). While he doesn't speak specifically about the problems facing Earth, he emphasizes the power of virtue to effect real change: Love, after all, is perhaps the strongest tie uniting a community. Unlike the approach to climate change in many other texts, *The Highest Frontier* is not afraid to engage with religious interpretations of the universe. As Hulme has also done, the novel suggests that older, humanist, and theological viewpoints might offer solutions to the problems of the Anthropocene. Certainly, the cultivation of virtue is something each of us, as individuals, can begin doing immediately. In doing so, perhaps we can achieve the necessary dispositions to begin forming a global community. At Frontera College, students like Jenny strive to become more human, offering us an example of our role in the Anthropocene.

The "Highest Frontier"

Another potential solution to Anthropocenic challenges is embedded in the novel's location at the "highest frontier" of high Earth orbit. President Dylan frequently draws attention to the dual metaphor of the college's name: "[N]o matter how perfect our [space] habitat, it remains truly a *frontera*; a frontier in outer space, just as Frontera College is a frontier in humankind's search for knowledge" (372). As the first college in orbit, Frontera represents a necessary alternative to Earth, with its "methane quakes, death belts, and invading ultraphytes" (15). Throughout the novel, the Unity party confronts the Centrist belief in the firmament by arguing that humanity needs to expand outward: Jupiter, for example, offers a "boundless source of fuel" (73). This aspect of *The Highest Frontier* suggests that space exploration might be the only long-term solution to the problems on Earth. Grinspoon similarly argues "we ultimately face a choice between spaceflight and extinction" (235). Nonetheless, *The Highest Frontier* does not focus on this bleak outlook on the future of Earth; rather, outer space offers tangible answers to Earth's problems. Ultraphytes, the invaders from "off-world," are an example of this kind of solution. As Jenny asks earlier in the novel, albeit uncomfortably, "What if ultra is…just the edge of something bigger?" (85). The novel hints that other aliens, other entities from outer space, may be just beyond the spacehab. By the end, however, as the college and Homeworld Security conclude that ultra have become an asset to Earth, we can read Jenny's question with more ease of mind. If there is more life in the universe—perhaps even intelligent life—they may have something valuable to offer Earth; they may even help humans understand themselves.

Grinspoon (2016) also advocates for this kind of attitude toward alien life; in his estimation, the kind of extraterrestrial species capable of contacting humans would be the kind that has achieved "planetary intelligence" (326). Because of the time and dedication reaching another planet would entail, the kind of beings capable of contacting Earth would have "the ability to act coherently and intentionally, on a global scale, on projects that persist for many millennia" (325). Humans could therefore learn from their example. Space exploration and the search for extraterrestrial intelligence are inherently valuable endeavors because they foster that kind of planetary intelligence. To implement successful scientific research and innovation, we have to learn how to work together more successfully. In a way, then, the search for extraterrestrial intelligence and the Anthropocene both call for a global mindset, a wise Earth. Ultimately, the efficacy of space exploration lies in its potential to offer us self-knowledge: The "kind of society that will thrive sustainably on Earth is one that embraces space technology for wise stewardship, for Earth observations, for asteroid deflection, for continued planetary exploration and the Earth wisdom it brings, and eventually for resources that will allow us to stop depleting our home planet" (Grinspoon 2016, 235). When the presidential candidates, through the influence of the wisdom plants, question the existence of the "firmament," *The Highest Frontier* similarly implies that space exploration is about growing "beyond what we are now" (375).

At the end of *The Highest Frontier*, Jenny discovers that her ultraphyte roommate has taken on the shape of a plant—fittingly, one of Jenny's wisdom plants. Using her Babynet diad, Mary proclaims, "**WISDOM WE ARE HERE**" (443). The conclusion reminds us of an earlier moment, when Jenny wishes she could just be a plant (142). While Jenny's wish stems from her desire to escape the stress of college life, and while Mary is not actually a plant, the plant theme might also be read as a metaphor. Plants, nature—we humans usually perceive these objects as separate, as "other" from humans. However, we are not separate from nature. Perhaps Mary is not really a plant, but she has come to identify with one; her last text even suggests that she's only come to understand and adopt wisdom now that she's a plant. Her last action in the novel sets yet another example for humans, the example of identifying with nature. Subtly, then, *The Highest Frontier* suggests a Gaian outlook. At the same time, however, it implies our formidable responsibility. After all, Jenny cannot escape her obligations simply by becoming a plant. In the end, she returns to Frontera College, despite its flaws, to continue pursuing wisdom, and to watch over her ultraphyte roommate.

Works Cited

Alcaraz, Jose M., et al. 2016. Cosmopolitanism or Globalization: The Anthropocene Turn. *Society and Business Review* 11 (3): 313–32.

Biermann, Frank. 2012. Planetary Boundaries and Earth System Governance: Exploring the Links. *Ecological Economics* 81 (4): 4–9.

Clarke, Bruce. 2017. Rethinking Gaia: Stengers, Latour, Margulis. *Theory, Culture & Society* 34 (4 July): 3–26.

Crutzen, Paul J. 2006. The "Anthropocene." In *Earth System Science in the Anthropocene*, ed. Eckart Ehlers and Thomas Krafft, 13–18. Berlin: Springer.

Crutzen, Paul J., and Christian Schwagerl. 2011. Living in the Anthropocene: Toward a New Global Ethos. *Yale Environment 360*, January 24. https://e360.yale.edu/features/living_in_the_anthropocene_toward_a_new_global_ethos.

Ernston, Henrik, and Erik Swyngedouw. 2018. Interrupting the AnthropoobScene: Immuno-biopolitics and Depoliticizing Ontologies in the Anthropocene. *Theory, Culture & Society* 35 (6): 3–30.

Grinspoon, David. 2016. *Earth in Human Hands: Shaping Our Planet's Future*. New York: Grand Central.

Haraway, Donna. 1991. A Cyborg Manifesto: Science, Technology and Socialist-Feminism in the Late Twentieth Century. Rpt. Donna Haraway. *Manifestly Haraway*, preface by Cary Wolfe, 1–90. Minneapolis: University of Minnesota Press, 2016.

Hulme, Mike. 2014. Climate Change and Virtue: An Apologetic. *Humanities* 3 (3): 299–312.

Kymlicka, Will. 2015. Solidarity in Diverse Societies: Beyond Neoliberal Multiculturalism and Welfare Chauvinism. *Comparative Migration Studies* 3: 17. https://doi.org/10.1186/s40878-015-0017-4.

Latour, Bruno. 2017. Why Gaia Is Not a God of Totality. *Theory, Culture & Society* 34 (2–3): 61–81.

Lovbrand, Eva, et al. 2015. Who Speaks for the Future of Earth? How Critical Social Science Can Extend the Conversation on the Anthropocene. *Global Environmental Change* 31: 211–18.

Peterson, Jordan B. 2019. Higher Education and Our Cultural Inflection Point. *Jordan Peterson Podcast* 21, February. https://jordanbpeterson.com/podcast/episode-62/.

Slonczewski, Joan. 2011. *The Highest Frontier*. New York: Tor.

———. 2019. "Re: Highest Frontier Question." Received by Alexa Dodd, March 7.

Yusoff, Kathryn. 2016. Anthropogenesis: Origins and Endings in the Anthropocene. *Theory, Culture & Society* 33 (2): 3–28.

Index

© The Editor(s) (if applicable) and The Author(s), under exclusive license to Springer Nature Switzerland AG 2020

B. Clarke (ed.), *Posthuman Biopolitics*,
Palgrave Studies in Science and Popular Culture,
https://doi.org/10.1007/978-3-030-36486-1

9783030364885